6/2012

Why Geology Matters

*The publisher gratefully acknowledges the
generous support of the General Endowment Fund of
the University of California Press Foundation.*

Why Geology Matters

Decoding the Past,
Anticipating the Future

Doug Macdougall

UNIVERSITY OF CALIFORNIA PRESS

Berkeley   Los Angeles   London

University of California Press, one of the most distin-
guished university presses in the United States, enriches
lives around the world by advancing scholarship in the
humanities, social sciences, and natural sciences. Its
activities are supported by the UC Press Foundation and
by philanthropic contributions from individuals and insti-
tutions. For more information, visit www.ucpress.edu.

University of California Press
Berkeley and Los Angeles, California

University of California Press, Ltd.
London, England

Library of Congress Cataloging-in-Publication Data

Macdougall, J. D., 1944–
    Why geology matters : decoding the past, anticipating
the future / Doug Macdougall. — 1
        p.   cm.
    Includes bibliographical references and index.
    ISBN 978-0-520-26642-1 (hardback)
    1. Historical geology.   I. Title.
    QE28.3.M334   2011
    551.7—dc22   2010043748

Manufactured in the United States of America

20   19   18   17   16   15   14   13   12   11
10   9   8   7   6   5   4   3   2   1

This book is printed on Cascades Enviro 100, a 100% post
consumer waste, recycled, de-inked fiber. FSC recycled
certified and processed chlorine free. It is acid free,
Ecologo certified, and manufactured by BioGas energy.

For Sheila, as always

CONTENTS

ILLUSTRATIONS

PREFACE

> Tho earth is not a mere fragment of dead history, stratum upon
> stratum like the leaves of a book, to be studied by geologists
> and antiquaries chiefly, but living poetry like the leaves of a
> tree, which precede flowers and fruit—not a fossil earth, but a
> living earth. Henry David Thoreau, *Walden*

In this excerpt from *Walden,* Thoreau inadvertently touched on something geologists and other scientists who study the Earth have appreciated for a long time but that has become increasingly important for
understanding our planet in recent years: the Earth, far from being
static, is dynamic and ever changing. Not a living Earth, exactly, but
an Earth with different parts that continually interact in ways that have
produced monumental changes over its long history. One way to read
that history is through Thoreau's strata, "like the leaves of a book"; there
are other ways, too, as will become apparent in the chapters that follow.

"History repeats itself" is an adage usually invoked to remind us that
by studying history we may be able to avoid mistakes of the past. That
may even be true, at least some of the time. Yet there are those—like
Nassim Taleb, author of *The Black Swan*—who argue just the opposite,
that history is not a good guide to the future and that the world is shaped
by events that are rare or unprecedented and therefore largely unpre-
dictable. Still, even proponents of this point of view don't advocate that

we all become fatalists, just that we learn to expect the unexpected. But if that is so, should we study history at all?

For the Earth, that is an easy question to answer, because only by understanding our planet's past can we anticipate its future. It is true that Nature can confront us with the unexpected: a "freak" storm, a devastating earthquake or tsunami, an asteroid impact. But these are only unexpected because they are rare in human experience. They are all things that have happened repeatedly during the Earth's history, and they obey the natural laws of physics and chemistry. That is why decoding the Earth's past is so important: the same physical and chemical principles that have governed our planet since its formation will also apply in the future. History really will repeat itself—if not in precise detail, then at least in general terms. Geology students are often taught that "the present is the key to the past." But earth scientists also recognize that in many ways the past is the key to the future.

The span of the Earth's history is immense: 4.5 billion years. It is hard to comprehend such a vast stretch of time, but with so much time available, even geological processes that operate at a snail's pace—like erosion of a mountain or movement of a tectonic plate—will eventually accomplish equally vast changes to our planet. It will take millions of years, but the Alps will be worn down to a flat plain, and Los Angeles will one day slide past San Francisco, heading north along the San Andreas Fault.

Such forecasts are interesting to contemplate, but they obviously don't have an immediate impact. No one reading this, or even their great-grandchildren, will witness these things. We are far more likely to experience the effects of some of the geoscience-related issues we hear about almost daily: floods, hurricanes, volcanic eruptions, earthquakes, deforestation, endangered species, mineral shortages—the list goes on, and I haven't even mentioned the big one, climate change. The problem is, most people without a background in the geosciences don't have a clear understanding of how Earth processes that affect our daily lives and our futures work, and this is true even for many of those who must manage society's response to these kinds of issues. Unfortunately, earth science subjects are usually given short shrift in the educational system. Recent data from the American Geological Institute, for example, show that in Texas—a state where the geosciences play an important role in the economy—only 2 percent of ninth-grade students are enrolled in earth science courses. For biology, the figure is 95 percent. My hope is that this book will, in some small way, stimulate a deeper interest in the field among its readers, and perhaps help to improve this situation.

Because knowledge of the Earth's past is such an important part of understanding how our planet works today and will in the future, this book, in addition to dealing with specific topics such as earthquakes and asteroid impacts, examines in a chronological way some of the important events that have affected the Earth over the past 4.5 billion years. Chapters exploring the Earth's history are interspersed with chapters devoted to specific events and processes; I hope that this structure, while a bit unconventional, proves both interesting and stimulating. As much as possible, I have tried to avoid getting bogged down in complex scientific discussions while still adhering to the main messages of the underlying research.

The twenty-first century is an exciting time to be a geoscientist. The stereotypical geologist trudging up a mountain with a pick still exists—and plays an important on-the-ground role. But the pick is now supplemented with an array of tools that make it possible to study the Earth and the solar system in unprecedented detail: satellites, supercomputers, electron microscopes, mass spectrometers, wireless communication, submersibles, and spacecraft, to name just a few. A geoscientist, Harrison Schmitt, was among the Apollo astronauts who visited the Moon (and brought back rock samples). Geoscientists regularly descend miles under the sea in research submersibles, don protective clothing to sample the gases escaping from active volcanoes, send off autonomous drifters and gliders to roam the oceans collecting data, or brave the rigors of the Antarctic to collect ice cores holding records of past climates. Some work in clean labs with samples so small that a single dust particle or fingerprint can contaminate an analysis; others feed their collected data into supercomputers to produce three-dimensional visualizations of parts of the Earth. Many of these scientists would not automatically think of themselves as geologists—their training is as likely to be in mathematics or oceanography as geology. But they are all part of an interdisciplinary effort to understand our planet's past, present, and future. It is in that sense that the word *geology* is used in the title of this book: geology is the study of the Earth, broadly construed.

The impetus for writing this book came partly from a desire to share some of the excitement about what geoscientists have learned about our amazing planet in recent decades. It also came from my conviction that, in a crucial way, earth science research really does matter for our future. Managing water, mineral, and energy resources sustainably, protecting biodiversity, planning for climate change and geological hazards—all of these things depend upon understanding how the Earth works as a system, and how it has responded to different conditions in the past. I hope that this book will bring that message home, and at the same time provide a glimpse of how earth scientists go about gaining that understanding.

ACKNOWLEDGMENTS

Many people helped to shape this book. I am especially grateful to two reviewers of the initial manuscript, Professors Richard Cowen and Ernest Zebrowski, who together made many valuable suggestions, identified errors, and generally pointed out ways to make the book more appealing. At University of California Press, Blake Edgar and Dore Brown worked hard to insure that the manuscript met their high editorial standards. Any problems that remain are entirely of my own making.

My agent, Rick Balkin, provided his usual invaluable support and perceptive suggestions throughout, from conception to final manuscript. Several people kindly made photographs available, which, I am sure readers will agree, enhance the text considerably: Abigail Allwood, Jim and Rebecca Brune, and Rick Otto. I'm also grateful to both the U.S. Geological Survey and NASA for their "open use" policy; several of the images used in the book come from these organizations. Thanks also to Ron Blakey for permission to use his paleomaps as a basis for several illustrations, and to Sigfus Johnsen and the Centre for Ice and Climate at the Niels Bohr Institute for permission to reproduce a diagram from Willi Dansgaard's *Frozen Annals*.

CHAPTER ONE

Set in Stone

In 1969, when I was a student in California, there was a rash of predictions from astrologers, clairvoyants, and evangelists that there would be a devastating earthquake and the entire state—or at least a large part of it—would fall into the ocean. The seers claimed this would happen during April, although they were not in agreement about the precise date. A few people took the news very seriously, sold their houses, and moved elsewhere. Others, a bit less cautious, simply sought out high ground on April 4, the date of the Big One according to several of the predictors. Cartoonists and newspaper columnists had a field day poking fun at the earthquake scare, and for us geology students the hubbub was amusing but also seemed a bit bizarre. Police and fire stations, along with university geology departments, got thousands of anxious telephone calls from nervous citizens. Ronald Reagan, then the state's governor, had to explain that his out-of-state vacation that month had been planned long in advance and had nothing to do with earthquakes. The mayor of San Francisco planned an anti-earthquake party for April 18, the sixty-third anniversary of the great 1906 San Francisco earthquake. He assured the public that it would be held on dry ground.

California didn't fall into the sea in 1969, of course, nor was there a huge earthquake (although there *were* earthquakes, as there are every year, most of them quite small). Astrologers can't predict earthquakes (or much else). Even earth scientists, with the best geological information and most up-to-date instrumentation, find precise earthquake prediction elusive, as we will see later in this book. However, the prognosis is much better for many other geological phenomena. And at the core of this geological prediction lies the kind of work geologists have traditionally done: decoding the past.

But how, exactly, do they do that? Where do earth scientists look to find clues to the details of our planet's history, and how do they interpret them? Those questions are at the heart of this book, and the answers are hinted at in the title of this chapter: the clues are found, for the most part, in the stones at the Earth's surface. (There are also many other natural archives of Earth history, such as tree rings and Antarctic ice. Ice cores in particular provide invaluable information about past climates. But these other records tell us only about the relatively recent geological past. Rocks allow us to probe back billions of years.)

To the uninitiated a rock is just a rock, a hard, inanimate object to kick down the road or throw into a pond. Look a little closer and ask the right questions, however, and it becomes more—sometimes much more. Every single rock on the Earth's surface has a story to tell. How did the rock form? *When* did it form? What is it made of? What is its history? How did it get here, and where did it come from? Why is this kind of rock common in one region and not in another? For a long time in the predominantly Christian countries of the West, answers to questions like these were constrained by religion. The biblical flood was thought to have been especially important in shaping the present-day landscape, and explanations for many geological features had to be built around the presumed reality of this event. However, as the ideas of the Enlightenment took hold during the seventeenth and eighteenth centuries, and as close observation of the natural world became ever more crucial for those seeking to understand the Earth, the sway of religion diminished and more rational explanations began to emerge. For geology especially, a field with its roots in the search for and extraction of mineral resources from the Earth, the pressure of commerce was also important. Those with the best understanding of how gold veins formed, or with the best knowledge of the kinds of geological settings likely to contain such veins, had the best chance of finding the next gold mine.

I will not dwell at length here on the history of geology's development as a science, or on the details of how early geological ideas evolved; these things have been dealt with in many other books. But it is worth pointing out a few key early concepts that revolutionized the way everyone—not just scientists—thought about our planet. Most of these intellectual breakthroughs arose in Europe (especially in Britain) in the eighteenth and early nineteenth centuries, and although there had been independent thinkers in the Middle East and elsewhere who had arrived at similar conclusions much earlier, the European versions would form the bedrock(!) of the emerging field of earth science.

What were these ideas and how did they come about? Without exception they stemmed from examination of rock outcroppings in the field together with observations of ongoing geological processes. One of the new concepts was that different rock types have quite different origins, something that seems obvious enough to us today. But in the eighteenth century a popular concept was that all rocks were formed by precipitation, either from a primordial global ocean or from the waters of the biblical flood. Those who championed this idea were dubbed—for obvious reasons—Neptunists, and they did not give up their theory easily. However, observations like those of Scottish geologist James Hutton, who described outcrops showing clear evidence that some rocks had once been molten, eventually turned the tables. The rock outcrops told Hutton a vivid story: flowing liquid material, now solid rock, had intruded into, and disrupted and heated up, preexisting rock strata. Hutton's descriptions of these once-molten rocks—not to mention the presence of active volcanoes like Vesuvius and Etna in southern Europe—led to the realization that there must be reservoirs of great heat in the planet's interior.

A second important early concept was that slow, inexorable geological processes that can readily be observed (rainwater dissolving rocks, rivers cutting valleys, sedimentary particles settling to the seafloor) follow the laws of physics and chemistry. Once again this seems an obvious conclusion in hindsight, but its implication—this was the revolutionary part for early geologists—was that geological processes in the distant past must have followed these very same laws. This meant that the physical and chemical characteristics of ancient rocks could be interpreted by observing present-day processes. Charles Lyell, the foremost British geologist of his day, promoted this idea as a way of understanding the Earth's history in his best-selling book *Principles of Geology,* first published in 1830. (The book was so popular it went through numerous editions and is still in print today in the Penguin Classics series.) Lyell himself was not the originator of the concept, but he called it the "principle of uniformitarianism," and the name stuck. Although the phrase itself is no longer in vogue, generations of geology students have learned that it really means "the present is the key to the past." And although the early geologists were primarily interested in working out the Earth's history, Lyell's principle of uniformitarianism can also be turned around: using the same logic, it is true that—to a degree—the past is a key to the future.

Finally, the most revolutionary of the new concepts was that the Earth is extremely old. This flew in the face of both the conventional wisdom of contemporary scholars and the religious dogma of the day. Once again, as with so much early geological thought, the idea of an ancient Earth was formalized by James Hutton, who wrote (in a much-quoted pronouncement about geological time), "we find no vestige of a beginning, no prospect of an end." No single observation led Hutton to the concept of a very old Earth; it was instead a conclusion he drew from a synthesis of all his examinations of geological processes and natural rock outcroppings—observations of things like great thicknesses of rock strata made up of individual sedimentary particles that could only have accumulated slowly, grain by grain, over unimaginably long periods of time.

With a foundation built on these new ideas, which were popularized and widely disseminated through Lyell's book, and with an ever-increasing demand for minerals and resources from the Earth, geology, now mostly free of religious fetters, exploded as a science during the nineteenth century. Countries developed geological surveys to map the terrain and discover resources, and universities founded departments of geological sciences. Decoding the past became a full-time occupation for a legion of geologists.

Today geology is subsumed into the much broader field of earth science, which includes everything from oceanography to mineralogy and environmental science. In a modern university earth science department, it is not uncommon to find researchers in the same building probing subjects as diverse as climate change, biological evolution, the chemical makeup of the Earth's interior, and even the origin of the Moon.

Let's return, however, to those clues to the Earth's past that are inherent in the physical and chemical properties of the planet's rocks—the clues that are set in stone. The challenge for earth scientists is to find ways to extract and interpret them, and in recent years very sophisticated techniques have been developed to do this. Nevertheless, there are also some very simple examples, long used by geologists, that illustrate how the approach works. Take the igneous rocks, those that form from molten material welling up from the Earth's interior. They come in many flavors, from common varieties familiar to most people, like granite or basalt, to more exotic types you may never have heard of, with names like *lamprophyre* and *charnockite*. The chemical compositions of these rocks can provide information about how they originated, but chemical analysis requires sophisticated equipment. On the other hand, there is a very simple feature—one that can be assessed quickly by anyone—that provides evidence about *where* the rocks formed. That characteristic is grain size.

Igneous rocks are made up of millions of tiny, intergrown mineral grains that crystallized as the liquid rock cooled down. How big these grains grow depends crucially on how fast the rock cools; lava flows that erupt on the Earth's surface cool rapidly, and the resulting rocks are very fine-grained. But not all lava makes it to the surface. Some remains in the volcanic conduits, perhaps miles deep in the ground. Well insulated by the overlying rocks, it takes this material a long time to cool, and the slowly growing mineral grains get much bigger than their surface equivalents. For this reason, rocks with exactly the same chemical makeup can have contrasting textures and *look* very different, depending on how quickly they congeal. This simple characteristic can be used to say something about the depth in the Earth at which the rocks formed.

Less obvious characteristics require more ingenuity to decode, but because the payoff—in terms of what can be learned about the Earth's history—is so great, earth scientists are continually searching for new ways to probe rocks. As we will see in later chapters of this book, geochemistry, especially the fine details of a rock's or an ice core's chemical composition, has become especially important. The behavior of chemical elements such as iron, or sulfur, or molybdenum, for example, depends on the amount of oxygen in their environment. As a result, the minerals formed by these elements are sensitive indicators of oxygen levels when they formed—and in some cases can be used to determine the amount of oxygen in the ancient ocean or atmosphere.

Similarly, analysis of isotopes has become one of the most important ways to extract information about the Earth's past. (Isotopes are slightly different forms of a particular chemical element; almost every element in the periodic table has several isotopes.) Often the conditions that prevailed when a sample formed can be deduced by measuring the abundances of different isotopes of a particular chemical element; we will encounter many examples of this approach in later chapters of this book. In an ice core, for example, oxygen or hydrogen isotope abundances might tell us about the temperature 100,000 years ago; in an ancient rock, isotopes might fingerprint the process that formed the rock, and allow us to investigate how similar or different that process was to those that occur today.

The very first application of isotopes in the earth sciences—aside from the use of radioactive isotopes for dating—still evokes admiration among geochemists and sometimes amazement from those who know nothing about geochemistry. It is a good illustration of how ordinary rocks can be a treasure trove of information about the past when the right questions are asked. In the late 1940s Harold Urey, a Nobel Prize–winning chemist at the University of Chicago, discovered from theoretical considerations that in some compounds the proportions of the different isotopes of oxygen depend on the temperature when the compound formed. In a flash of insight, he realized that this property could be used to deduce the temperature of the ancient ocean—a groundbreaking idea. Urey proposed that measurements of oxygen isotopes in the calcium carbonate shells of fossil marine organisms could be used to calculate the water temperature when these creatures grew. He and his students then verified the theory by making those measurements, and in doing so they pioneered the field of "paleotemperature" analysis. Since that early work, tens of thousands, if not hundreds of thousands, of oxygen isotope measurements have been made to document in fine detail how seawater temperatures have fluctuated in the past. In my humble opinion—perhaps with a slight bias because my own background is in geochemistry—Urey's paleotemperature work ranks among the all-time great advances in the earth sciences.

Different rock types raise different questions about the past, of course, or at least allow different questions to be asked, but well-defined approaches for extracting evidence have been worked out by earth scientists for most rock varieties within the three great categories: igneous, sedimentary, and metamorphic. These familiar subdivisions of the rock kingdom are based on mode of formation: igneous rocks such as granite are formed from molten precursors, as James Hutton was one of the first to realize; sedimentary rocks result from the deposition or precipitation of particles, usually from water; and metamorphic rocks arise when any preexisting rock is changed chemically and/or physically, typically when heated or stressed during a process like deep burial or mountain building. Current theories about how the outer part of the Earth formed and has evolved rest on evidence derived mainly from the chemical properties of igneous and metamorphic rocks, which are the primary components of both the continents and the seafloor. But in many ways sedimentary rocks are the most important for decoding the Earth's history.

Why should that be? There are at least two reasons. First, they form at the Earth's surface, mostly in the sea but sometimes (as in the case of sandstones composed of desert sand) in contact with the atmosphere. This means that, potentially, these rocks incorporate information about the Earth's surface environment in the distant past. And second, many sedimentary rocks contain fossils, the primary record of how life on Earth arose and evolved. Without fossils, our understanding of evolution would be rudimentary.

By putting together thousands upon thousands of stories from studies of individual igneous, sedimentary, and metamorphic rocks and rock outcroppings, earth scientists have gradually woven together a history of the Earth. As for most histories, the details become less sharp the farther back one probes. Some of the most ancient evidence is missing entirely, or made difficult to interpret because geological processes operating over the Earth's long history have altered the rocks' characteristics and muddled the clues they contain. Nevertheless the narrative of our planet's evolution as we know it today is a superb scientific achievement. It is also a story in revision, continually updated as new discoveries are made and improvements in analytical capabilities allow new questions to be asked.

But what about chronology? How have earth scientists determined the timescale of this narrative? Events need to be ordered in time if we are to understand their significance; it isn't very helpful to know the temperature of the seawater in which a fossil animal grew if you have no idea *when* it lived. Ever since Hutton's "no vestige of a beginning, no prospect of an end"—and even before that—earth scientists have sought ways to determine the ages of rocks and the Earth as a whole.

The ultimate goal—the development of techniques that could provide the "absolute" age of rocks in years—came within reach only with the discovery of radioactivity near the end of the nineteenth century. We will come to that shortly. But long before radioisotope dating methods were devised, earth scientists had already developed early versions of the geological timescale, placing important events from the Earth's history in a time sequence (see figure 1 for a modern version; if you are not already familiar with the names of geological eons, periods, etc., you may want to refer to this figure repeatedly as you read this book). How did they do this?

As early as the 1660s Nicolas Steno, a Danish anatomist who had an insatiable curiosity about the natural world, realized that rocks at the bottom of a stack of sedimentary layers must be older than those at the top. Steno was living in Italy at the time, and his observations were made while he was examining sedimentary rocks in the Alps. His insight was that the Alpine sedimentary strata—and the fossils they contain—have time significance. It is only relative time significance, to be sure; Steno could say whether a particular layer was older or younger than neighboring layers, but he couldn't determine its actual age. All this may seem obvious now, but at the time it was a breakthrough. By studying the inert rock layers of the Alps, Steno was able to visualize the nature and timing of their formation. Today he is generally regarded as the founder of the field of stratigraphy, the scientific study of sedimentary rock strata.

From Steno's time onward, his simple principle of ordering sedimentary layers in time was used to work out the relative chronology of geological events. This was easy enough to do in local areas where distinctive individual layers could often be traced from one rock outcrop to another. But long-distance correlation was difficult. Was a limestone layer in France the same age as one in England or Sweden, or across the Atlantic in the United States? It was difficult to say. Regional relative timescales could be constructed, but a global one seemed beyond reach.

However, there were clues in the sedimentary rocks that helped

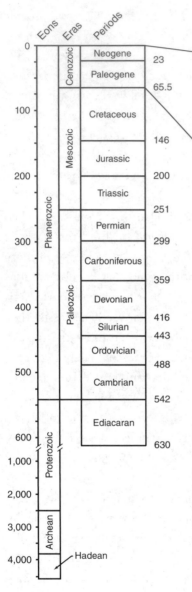
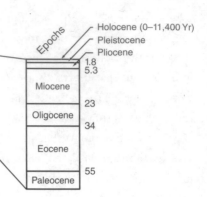

Figure 1. The geological timescale, with dates in millions of years before the present. Note the change of scale at 700 million years, during the Proterozoic eon. Only the major subdivisions of the timescale are illustrated here; geologists have defined many finer intervals within these. (Based on the most recent data from the International Commission on Stratigraphy.)

resolve this dilemma. Long before Charles Darwin wrote about evolution, earth scientists recognized that life on Earth had changed through time. Wherever they looked, they found the same story. Fossils in the youngest rocks near the top of sedimentary sequences looked similar to living forms, but in lower, older layers, the fossils were often small and quite different from any known plants or animals. And in some places, below (and therefore even older than) the rocks containing the old, unfamiliar fossils were strata completely barren of any sign of animal or plant life.

An English surveyor named William Smith was one of the first to recognize the practical significance of this changing sequence. Surveying was his trade but geology was his passion, and as he traveled around the British Isles in pursuit of his profession he took notes about the local geology and collected fossils. He noticed that the sequence of fossils, the way the assemblages of organisms changed as he proceeded from older to younger rocks, was always the same—even if the rocks themselves looked quite different. Half a century or more before Darwin published his *Origin of Species,* Smith organized his fossil collection, which he proudly displayed to friends, according to relative age, not in groups of similar-looking organisms as most contemporary collectors did. Although he didn't know it, he was using evolution, as recorded by the fossils, as a way to make correlations among sedimentary rocks formed at the same time but in far-flung localities. The goal of a global relative timescale was a step closer.

Those who followed in the footsteps of Steno and Smith gradually built up the geological timescale until they had filled in most of the subdivisions shown in figure 1, from the Cambrian period to the present. The names they gave to the major subdivisions of this timescale, particularly the names of the geological periods, usually referred to geographical regions where fossil-containing rocks of that particular time were abundant and first described in detail—for example, *Jurassic* after the Jura Mountains of Switzerland, or *Ordovician* and *Silurian* after two ancient tribes that lived in different parts of Wales. All of this was done before the discovery of radioactivity, and there was no real sense of the great span of time represented. And because the relative timescale was based on fossils, it was blank below the base of the Cambrian period. As far as early geologists could tell, rocks older than this did not contain any fossils at all (as we will see, there *was* life on Earth long before then, but most of the fossils from those earlier times are rare, small, and easy to overlook). The ancient, apparently barren rocks were simply referred to as "Precambrian."

This early relative timescale was in reality a record of the evolution of marine life. Although there were geographical variations in life forms in the past, just as there are today, the general pattern of evolution is clear enough in the fossil record that sedimentary rocks anywhere in the world could be placed in the correct sequence, as long as they contained fossils. Devonian rocks in Europe, for example, contain fossil assemblages that are recognizably similar to those in Devonian rocks from America or Africa. This helped greatly in the construction of the timescale because there is no single locality on Earth where rocks spanning the entire time from the Cambrian period to the present—or even a significant portion of it—occur in a continuous, uninterrupted sequence of sedimentary layers. The timescale had to be constructed bit by bit through detailed examinations of small portions of the geological column (as it is often called) in different places, coupled with correlation between localities where there was obvious overlap. This might seem at first to be an ad hoc approach, but it has been extremely successful, as the timescale in figure 1 attests. So complete is our understanding of evolution that an experienced field geologist can walk up to an outcrop of sedimentary strata anywhere in the world and, if he can find a few fossils, place it quite precisely in the geological timescale.

All of this has been accomplished in spite of the fact that only a very small fraction of all species that have ever existed on Earth occur as fossils. It is simply not very easy to become a fossil. Most estimates suggest that fewer than 1 percent of species have been preserved in rocks, and it is easy to understand why. Even in the most favorable environments—a quiet sea bottom with slowly accumulating muddy sediments, for example—most dead organisms are consumed by scavengers or simply rot and dissolve away before they can be preserved. Usually only the hard parts—shells, bones or teeth—are preserved, and even then it may be only a fragment. Adding to the challenge is the fact that it is sometimes difficult to deduce the whole from the parts, especially for unfamiliar organisms. Sharks' teeth are relatively common as fossils, but for a long time—in spite of the fact that sharks were well known— nobody knew what the fossils were because they were isolated objects, not obviously associated with anything else. And even if complete fossils are preserved, the sedimentary rocks that contain them may later be destroyed by erosion or metamorphism. Darwin was one of the many scientists concerned about the resulting gaps in the fossil record.

Nonetheless, even with the limited available sample of fossil species, sedimentary rocks have yielded up in amazing detail the story of how life on Earth has changed. The early geologists placed boundaries between individual eras and periods, and even between finer subdivisions of the timescale, at places in the geological column where they observed rapid changes in the types of fossils preserved. The names of the three eras shown in figure 1—*Paleozoic, Mesozoic,* and *Cenozoic*—are derived from Greek for "ancient life," "middle life," and "recent life," because of the abrupt and truly radical changes in fossil species that occur at the boundaries between them, with the preserved life forms becoming increasingly familiar toward the present. The boundaries can be readily identified everywhere on Earth where rocks of these ages occur, and we now know that they record short periods of widespread extinctions, when large fractions of the organisms inhabiting the oceans were wiped out through catastrophic environmental disruption. The extinctions were followed relatively quickly (in geological terms) by evolution and radiation of new life forms. Less drastic but still major changes in the nature of marine life mark the boundaries between the geological periods.

The timing of these changes was, for a long time, elusive. As the nineteenth century drew to a close, scientists of all stripes were working on ways to measure geological time. Physicists wanted to know the age of the Earth; geologists wanted to know the ages of individual rocks and the duration of different parts of the timescale. Many ingenious approaches were tried, but most of them rested on questionable assumptions and all had very large associated uncertainties. The most extreme estimates of the Earth's age were in the range of a few tens of millions of years up to perhaps 100 million years. There was simply no reliable way to know how much time was represented by Precambrian rocks, or to work out anything about the rate of evolution.

That all changed with the discovery of radioactivity in 1896. Once it was understood that radioactive isotopes decay at a constant rate, their potential for geological dating became clear. One of the early pioneers of radioactivity research, Ernest Rutherford, was the first to make that leap. He was a physicist and an experimentalist, and he asked his geologist colleagues to give him rocks they thought were very old. From measurements of the radioactive isotopes in these samples, he calculated that they were about 500 million years old. This was a startling result, and it shook up the scientific establishment. If Rutherford's result was accurate, it meant that the Earth as a whole was even older than 500 million years, and therefore much older than was generally thought.

By today's standards, Rutherford's experiments were crude. Geochronology, the science of dating rocks, has made huge advances over the century or so since he made his initial measurements. The approach is still the same, based on the knowledge that radioactive isotopes decay at a known, constant rate. But today's analytical instruments are capable of making very precise measurements on small amounts of material, and the dates that result are also very precise. All of the boundary ages shown in figure 1 are based on radiometric dating (as the process is usually called), and the same techniques have shown that the Earth is between 4.5 and 4.6 billion years old. Time is such an important part of decoding the past that it is worth spending a few pages examining just how radiometric dating works.

The first thing to say is that geological time is immense. Four and a half billion years is a very long time, hard to comprehend from a human viewpoint. In this era of billionaires and trillion-dollar bailouts, the number itself is not unusual, but its enormity becomes apparent when it is put into perspective. Our species, *Homo sapiens,* has been around for about 200,000 years or perhaps a bit less, a very long time by most standards. But that is a minuscule fraction, just a *few millionths,* of the Earth's age. A commonly used analogy is a hypothetical three-hour movie depicting the Earth's history. Three hours is very long for a movie, but even so *Homo sapiens* would appear only in the last half second

One of the implications of the enormous span of geological time is that even though many geological processes seem to operate at insignificantly slow speeds, they can wreak huge changes. Tectonic plates, as we will see in a later chapter, move at speeds of only a few inches per year, yet multiply that by hundreds of millions of years and whole new ocean basins can open up and then disappear again. Over similar timescales great mountain ranges can be thrust up, then worn down to a flat plain by erosion.

But to return to the details of the dating methods used to measure these great swaths of geological time: fortunately, there are many elements in the periodic table that have naturally occurring radioactive isotopes, and many natural materials contain small amounts of one or more of them. This means that in principle, and with judicious sample selection, almost anything can be dated. However, each of the dating procedures that has been developed has its own limitations. For example, radiocarbon dating, which is probably the most widely known of all the geochronological methods, can only be used to date organic material that was once part of a living plant or animal, and is also restricted to material younger than about fifty thousand years. This limited time span results from the fact that the method is based on the radioactive decay of the isotope carbon-14, which decays away very quickly. (Isotopes are labeled according to the number of neutrons plus protons in their nucleus—fourteen in the case of carbon-14. In scientific writing this is usually shown as a superscript to the chemical symbol, i.e., ^{14}C, but here I'll use the longer and easier to read format, i.e., carbon-14.)

The radiometric dating methods most commonly used for rocks employ isotopes of elements that are relatively abundant and familiar, like potassium and uranium, and also some that are more exotic, such as rubidium, rhenium, and samarium. Each of the methods has its own advantages and disadvantages, and often the circumstances—things like the geological setting of the sample to be dated—dictate which method is most likely to be useful. The most commonly used technique for ancient rocks, one that we will encounter repeatedly in this book, is based on the decay of uranium to isotopes of lead. One of the reasons uranium-lead dating is so useful is that a wide range of rock types contain a mineral that can be easily extracted for analysis and is both naturally rich in uranium and very resistant to alteration: the mineral zircon.

As you might guess from its name, zircon is rich in zirconium, and from its chemical formula, $ZrSiO_4$, it is apparent that silicon and oxygen are its other major constituents. Uranium is present only in trace quantities but still at much higher concentrations than in most other minerals, because uranium atoms easily take the place of zirconium in the mineral's structure. Zircon is a hard and dense mineral that is usually reddish in color; small grains of it are ubiquitous in igneous rocks. Rarer large crystals are sometimes sold as semiprecious gemstones, but for geologists, zircon's real value lies in its usefulness for dating. It is so resistant to alteration that even when rocks are buried, heated, and undergo significant metamorphism, the zircon crystals often remain unscathed—and retain the age of the original rock. When rocks like granite are weathered at the Earth's surface, many of the minerals they contain dissolve away or are turned to clay, but zircon crystals survive. Because of this, beach sands invariably contain grains of zircon.

Alongside uranium-lead dating, a second widely used radiometric dating technique that we will encounter involves the decay of potassium to an isotope of the gas argon. There is no potassium-rich equivalent of the mineral zircon, but because potassium is a relatively common element at the Earth's surface, many common minerals—for example, certain types of mica and feldspar—can be dated using this technique. For various technical reasons, potassium-argon dating is especially useful for the younger parts of the geological record.

Most of the dates for the modern geological timescale were measured using either uranium-lead or potassium-argon dating. In cases where the right samples were available, both methods were used; such cross-checking using independent techniques ensures that the results are accurate. But there is an issue concerning the ages shown in figure 1 that needs to be addressed: there are significant difficulties in applying both the potassium-argon and uranium-lead dating methods to sedimentary rocks, and as we saw earlier, fossils in sedimentary rocks are the basis of the timescale. How, then, were these ages obtained?

The problem becomes clear if we consider how sedimentary rocks are formed. Many of the mineral grains that comprise them were originally part of other rocks on the continents; they were eroded from their parent rocks, carried to the sea, and deposited. Dating these minerals would give the ages of the parent rocks, not the sedimentary rocks themselves. Furthermore, the minerals in sedimentary rocks that are directly precipitated from seawater (and thus would be appropriate for dating these rocks) don't contain enough uranium or potassium or other radioactive isotopes to make them useful for age determinations. Calcium carbonate, a widespread component of ocean sediments, falls into this category. This is one instance in which Mother Nature has not been very kind to earth scientists.

But obviously the problem has been circumvented; accurate radiometric dates do exist for many sedimentary rocks. What was the solution? The answer has to do with the fact that the Earth is a very active planet, with volcanoes spewing out volcanic ash almost continuously. This was brought home with a jolt early in 2010 when an ash cloud from the Eyjafjallajökull volcano in Iceland shut down air travel across Europe. (Pronouncing the name of this volcano may seem daunting to most native English speakers—but I understand that "I forgot the yogurt" is not a bad approximation.) The Eyjafjallajökull eruption was tiny by global standards, but it illustrated how ash from a single volcano can spread over a wide region. The largest eruptions disperse ash globally, and when this material settles to the seafloor it forms layers that are easily recognizable. The ash layers are markers of essentially instantaneous geological events, and are therefore ideal candidates for dating. Fortunately, they often contain zircon crystals, or minerals that can be dated using the potassium-argon method.

Volcanic ash layers are so abundant that they have become by far the most important material used for dating sedimentary rocks. The most explosive volcanoes—the ones that produce the most ash—occur mainly along the margins of ocean basins as a result of plate tectonic processes. Think of the volcanoes of Indonesia, or the Andes. Even if individual volcanoes erupt only sporadically, sediments accumulating in these regions are laced with ash layers. If it is important to know the age of a particular level in the sediments—say, a level that marks a geological boundary—and there doesn't happen to be a convenient ash layer in exactly the right position, it is usually possible to interpolate between closely spaced layers.

A case in point is a sequence of limestone beds in southern China that extends across the boundary between the Permian and Triassic periods. Fossils show that this boundary—which is also the divide between the Paleozoic and Mesozoic eras (figure 1)—marks the most extensive mass extinction event in the Earth's history, a time when more than 90 percent of the species living in the oceans quite suddenly disappeared. Accurate dating was a high priority, but the limestone couldn't be dated directly. Fortunately, though, it was deposited in a volcanically active region and contains numerous interbedded ash layers. In the 1990s a team of geochronologists from the Massachusetts Institute of Technology (MIT) sampled ash layers from above and below the boundary, painstakingly separated out the small zircon crystals they contain, and measured the zircon ages using uranium-lead dating. The

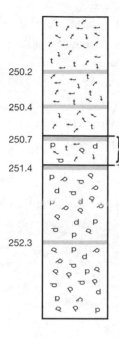

Figure 2. Sedimentary layers spanning the Permian-Triassic (P-T) boundary in southern China. The predominant rock type here is limestone; ash layers are shown as gray bands. Ages for the ash layers, based on uranium-lead dating of zircon crystals, are given in millions of years. The ash layer immediately below the formally defined position of the P-T boundary provides the age of the boundary. The locations of Permian ("p") and Triassic ("t") fossils are shown. Note that there is a small interval of mixed fossils immediately above the P-T boundary; this is a common feature of sedimentary rocks, and is caused by burrowing organisms and currents that stir the sediments as they are laid down. The column shown here represents just over sixty feet of rock layers. (Based on data in Bowring et al. 1998.)

results are shown in figure 2. They show that a segment of the sediment sequence several yards thick and spanning the boundary was deposited over a time period of just two million years. The dates for the ash layers also pin down the age of the boundary precisely to 251.4 million years. Perhaps equally important, by dating closely spaced ash layers these researchers were able to conclude that the great pulse of extinctions occurred over a short time span, less than a million years.

I have not yet said anything about the Precambrian part of the timescale. Radiometric dating has uncovered its true extent; the scientists who worked out the early versions of the timescale from fossils would be astounded to know that it comprises more than 85 percent of the Earth's history. Lacking fossils, rocks from the Precambrian can only be placed in a time context by direct radiometric dating. Figure 1 shows only a few major divisions of this part of the timescale: the Hadean, the Archean, and the Proterozoic. Ages for the boundaries between these subdivisions are partly arbitrary, and partly based on recognized events that have affected the Earth globally. In spite of the antiquity of the Precambrian rocks, they have revealed a rich tapestry of sometimes surprising information, as we will see in later chapters. They depict a world that for billions of years was very different from the one we know today.

This brief introduction is meant to provide an overview of how earth scientists use different types of information stored in rocks to decipher events from the Earth's past and to work out their chronology. That effort is ongoing, and new discoveries continually sharpen or modify different aspects of the story. In recent years special emphasis has been put on identifying times and events in the past that have relevance to what may occur in the future. This has become particularly important for issues that will affect the near-term future of human societies, such as global warming. Before we turn to such concerns, however, the following chapter goes back to the very beginning, 4.5 billion years ago, to explore our planet's origin and its very earliest days. We have no earthly rocks left over from that time; any that once existed have long since been destroyed by geological processes. What we do have, however, are rocks from space. Like Earth rocks, they too have stories to tell.

CHAPTER TWO

Building Our Planet

In 1969, the same year astrologists were predicting a big earthquake in California, another event was unfolding at the other end of the world that caught the attention of earth scientists. Its consequences were far-reaching. Japanese scientists working in Antarctica came across small, dark-colored rocks scattered across the surface of the ice. That might not seem unusual except that the scientists were working in a region that was completely blanketed in snow and ice; there was no obvious local source for the rocks. As it turned out, they were not from anywhere nearby—they had arrived from space. The rocks were meteorites.

That posed yet another puzzle. It has long been known that large quantities of extraterrestrial material fall on the Earth; most estimates put the amount at about one hundred tons per day. But most of that is in the form of tiny dust particles that burn up in the atmosphere—the "shooting stars" we see in the night sky—and only a very small fraction of the total arrives as recognizable meteorites. With two-thirds of the Earth covered by oceans, the majority of meteorites simply fall into the sea and are never recovered. Very few people have ever witnessed one landing, and you certainly don't find them lying around in your garden or the local park. So the numerous meteorites sitting on the Antarctic ice seemed to defy explanation. What was so special about the southern continent that it collected more meteorites than anywhere else on Earth?

As usual with initially puzzling phenomena, there was a rational explanation. It soon became clear that a fortuitous process was at work: meteorites that had fallen in the Antarctic over tens of thousands of years were being gathered up and delivered to small, concentrated areas known as "blue ice" regions (the Japanese scientists were working in one of these). The blue ice occurs in places where thick glaciers, slowly flowing outward from the continent's interior, encounter a buried topographical barrier such as a mountain range and are forced upward. As fast as the flowing ice reaches the surface, the constant high winds and dry air of the Antarctic ablate it away, and the meteorites carried within the glaciers are left as a deposit on the surface. The concentrations are similar to the so-called lag deposits often found in deserts, where wind blows away the finest grains and leaves behind a layer of coarse, heavy material. Like a gigantic conveyor belt, the Antarctic ice transports thousands of years of meteorite falls to a few small areas where the ice simply disappears and the space rocks remain. Since the 1969 discovery, regular expeditions have been organized during the Antarctic summer to collect meteorites from the blue ice regions. Scientists from around the globe descend on the continent to join the search. Over just a few decades, tens of thousands of new meteorite samples have been recovered, increasing the size of the world's collections many times over.

Why such an interest in meteorites? Because they are, many earth scientists will tell you, keys to understanding how the Earth and the solar system formed. Superficially many meteorites don't look significantly different from most Earth rocks. But close study shows that they are very different, and the differences hold clues to their distant origin. Some ancient cultures venerated meteorites because they were thought to have been sent by the gods; now they are venerated by scientists because they bring information from the earliest parts of solar system history. Nearly every meteorite dated by the radiometric techniques described in the previous chapter has an age close to 4.5 billion years, significantly older than the Earth's oldest rocks. And the most common type of meteorite—known collectively as the "chondrites"—contains clues about the kinds of material that went into the construction of our planet. Indeed, the mineral assemblages in some chondrites are probably a good representation of the ones in the rocks that were swept up to form the Earth.

The chondrites are just one of a variety of meteorite types, all ancient, but each with a different history—and each containing clues about how the Earth and other planets formed and evolved. For example, the members of one group, the iron meteorites, are made up of solid iron metal, alloyed with a modest amount of nickel. All the evidence indicates that these meteorites are analogs of the metallic cores that inhabit the interiors of planets (more on the Earth's iron core later in this chapter). If you have an opportunity to visit the American Museum of Natural History in New York, you can put your hand on a piece of one of these iron meteorites, probably once part of the core of a small asteroid. Dating to 4.5 billion years ago, it will undoubtedly be the oldest thing you have ever touched. This massive piece of iron metal, weighing about thirty-four tons, is from one of the largest meteorites known, the Cape York meteorite, named (as are all meteorites) after its finding place, in this case Cape York in northwestern Greenland. It was "discovered" in 1894 by the American explorer Robert Peary (Inuit people had known about it for centuries before, and had used it as a source of iron). Dating studies show that the Cape York meteorite fell to Earth about ten thousand years ago, breaking up into numerous pieces as it crashed through the atmosphere. The specimen at the Museum of Natural History is the largest of the many samples of this meteorite displayed in museums around the world.

In contrast to the iron meteorites, the chondrites are made up of a jumble of mineral grains, many of them familiar constituents of rocks on Earth, but also including iron metal—which does not occur in terrestrial rocks—and small, marblelike spherical objects called "chondrules" (it is from these that the chondrites take their name). The chaotic texture of chondrites indicates that they were formed in a process that randomly swept together their different components and cemented them together. The texture is also an instant clue to one of their most important characteristics: they have never been melted. Furthermore, mineral grains in the chondrites have provided the oldest ages ever measured by radiometric dating; they date from the very earliest days of the solar system. These two properties have led the geochemists who work on these intriguing objects to conclude that chondrites bring us something we cannot find on Earth or among other meteorite varieties: an unprocessed sample of the original material from which the Earth and our neighboring planets were made. They appear to be unaltered samples of the solid matter that was floating around in the solar system just as the Earth was being born; there is strong evidence that they come from small asteroids that never grew big enough to heat up and melt, as larger objects did.

The primitive nature of the chondrites has made them a cornerstone for information about the Earth's overall chemical composition. You might wonder why these rare rocks from space should play such a key role when the whole Earth is right below our feet, ready for us to analyze. You might also wonder why it is important to know the Earth's overall composition in the first place. The answer, in brief, is that if we want to understand how the Earth got to its present state, we have to know something about its initial, overall composition. That is not easy to discover when we only have access to rocks from our planet's thin, outermost skin—which is very different from the inaccessible interior. However, by using information from the chondrites as a kind of reality check, and integrating that data with direct measurements on surface rocks and information about the interior obtained using remote sensing methods, geochemists have been able to work out models for the Earth's overall composition that satisfy independent evidence such as the density of our planet.

An important concept in formulating these models is that the composition of the Earth and the other "terrestrial" planets (the solar system's inner, rocky planets, Mercury, Venus, and Mars) depends on the proportions of the main minerals found in the chondrites that each planet incorporated. A good example is iron, which is so abundant and so heavy that the well-known density variations among the terrestrial planets can be attributed almost entirely to differences in their iron contents. Grains of iron metal are abundant in the chondrites, but different chondrites contain different amounts. The reason the Earth is much denser than Mars, according to the models, is that its chondritelike building blocks happened to contain more iron. Similarly, other differences among the terrestrial planets can be understood in terms of different planets incorporating different amounts of the various constituents of chondrites. This is undoubtedly an overly simplistic description of what actually happened, but that doesn't invalidate the chondrites as a good starting point for understanding the composition of the Earth and other planets. The different planets just ended up incorporating different amounts of the various constituents of these meteorites.

But even if these meteorites give us a good understanding of the Earth's chemical composition, they don't tell us much about the *process* of planet formation. What happened 4.5 billion years ago that caused our planet to come into existence? A few clues come from meteorites, but most of the evidence derives from other sources, especially astronomical theory and observations.

In 1990 NASA launched the now-famous Hubble Space Telescope. Although there were early problems with its optics, these were eventually corrected and the instrument has since sent back clear, stunning images of other worlds deep in space. Some of the most arresting of these are pictures of enormous, towering clouds far out in what we usually think of as "empty" space. The clouds were well known to astronomers from ground-based telescopes before the launch of Hubble, but the space telescope images are especially beautiful. They depict huge, irregular, wispy, chaotic, and sometimes menacing-looking clouds made up of gas and dust. These awe-inspiring features, it turns out, are where new stars and planets are born.

Using telescopes like Hubble, astronomers have been able to observe star formation in interstellar clouds directly. Other kinds of astronomical observations show that many stars have planets orbiting around them. (These latter observations are difficult, and so far only planets far bigger than the Earth have been detected, but hundreds of them have been found since the first one was discovered in 1995. Most astronomers are confident that it is just a matter of time and improvements in technology before smaller, rocky, Earth-like planets are discovered orbiting distant stars.) All the evidence, then, suggests that solar systems like our own are not unusual, and that the material that now makes up our Sun and planets was once part of an interstellar dust-and-gas cloud like the ones imaged by Hubble. Direct observations show that the main components of the clouds are gaseous hydrogen (the most abundant chemical element in the Universe) and helium. But they contain other things too, including a wide variety of solid "dust" particles: microscopic crystals of frozen, hydrogen-rich compounds such as methane, ammonia, and ordinary water; tiny particles of clay; and grains of several other minerals that are common on Earth.

But how does one of these vast clouds get transformed into stars and planets? The first thing to note is that the interstellar clouds are not homogeneous; they are turbulent, and some regions are much more matter-rich than others. Either on their own or because of some external trigger, these denser regions begin to pull surrounding material into their centers through gravitational attraction. Once started, this process of gravitational collapse is self-sustaining; as the central region gets denser, its gravitational attraction increases. Soon a portion of the dispersed cloud has been transformed into a very dense and hot central body—the beginnings of a new star—surrounded by a spinning disk of cooler, leftover material that will eventually form planets. The interstellar clouds are so huge that multiple solar systems can form from just a part of them.

Clues found in some meteorites suggest that a nearby supernova explosion may have been the trigger for the gravitational collapse that led to the formation of our own solar system. Supernovae are exploding stars, and although astronomers recognize several different types of supernovae, all of them involve stars larger than the Sun. When a large star runs out of nuclear fuel in its core, it collapses catastrophically; its central region heats up to such a high temperature and is compressed to such an immense pressure that a sudden burst of nuclear fusion occurs, generating a gigantic explosion that literally rips the star apart and sends its contents hurtling out into space. The explosion also sends gigantic shock waves rippling outward, waves with the potential to compress already dense parts of an interstellar cloud and initiate the star- and planet-forming process.

Astronomers estimate that a supernova explosion occurs somewhere in the Universe roughly every second. Even if you read fairly quickly, half a dozen or more stars have exploded in the Universe since you started reading this paragraph. In spite of their abundance, however, the Universe is so vast that supernovae explosions in our own cosmic neighborhood are relatively rare. The last one close enough to be visible to the naked eye (it was nevertheless very far away) occurred in 1604. When they do appear, they suddenly light up in the sky, looking like a new star; they glow brightly for a few weeks or months, then gradually fade away. Chinese astronomers recorded sightings of supernovae (although they didn't know what they were) almost two thousand years ago.

The intense burst of energy that accompanies a supernova explosion initiates further nuclear reactions that create an array of radioactive isotopes. These, like other material from the exploding star, are spewed out into space. Nuclear physicists have calculated in great detail which isotopes are produced during these explosions, and in what quantities. Remarkably, mineral grains in some chondritic meteorites contain traces of these isotopes. These grains, which other evidence indicates grew in the disk around the nascent Sun, evidently captured material from a supernova as they formed. The specific radioactive isotopes they incorporated decay away so quickly that they would not have been present (they would have completely decayed away) if more than a few million years, at most, had elapsed between the explosion and formation of the minerals. The story that the meteorites tell is that a nearby supernova explosion injected its products into the dust and gas cloud just before the solar system began to form. It is possible, and even likely, that the shock wave spreading out from this explosion was the trigger for the gravitational collapse that quickly led to the formation of our Sun, the chondrites, the planets, and, eventually, to us.

Computer simulations indicate that once collapse begins in part of an interstellar cloud, formation of the flat, rotating disk of matter with a protostar—in our case the protosun—at its center is rapid. As the material surrounding a protostar gets compressed into a disk, it heats up to high temperatures—so high that most or all of the dust particles from the interstellar cloud are vaporized. These conclusions from theory are confirmed by observations: astronomers have detected disks of gas and solids surrounding forming stars, and studies of chondritic meteorites show that many of their mineral constituents precipitated from the hot vapor of the disk as it cooled. Only minuscule amounts of unaltered dust particles left over from the interstellar cloud have ever been identified in meteorites.

There are few observations to guide us through the processes that transformed the protosun and its surrounding hot disk into our present-day solar system; for the most part we have to rely on computer simulations. What follows is a very brief outline of how scientists think it may have happened. First (and this is one of the more certain parts of the story) the protosun sucked up nearly all surrounding material and became so hot and dense that nuclear fusion reactions ignited in its interior. That was the real birth of our Sun; the fusion of hydrogen (by far the most abundant element in the Sun) into helium was, and still is, the source of its energy. The Sun contains about 99.9 percent of all material in the solar system, but obviously the matter remaining in the surrounding disk was sufficient to create all the other inhabitants of today's solar system: planets, their various moons, asteroids, and comets.

As the disk around the young Sun cooled down, mineral grains started to precipitate from the vapor, and as the grains became more numerous they began to collide with one another in their orbits around the Sun. Small grains, computer simulations show, usually stick together when they collide, so the average size of objects in the disk increased rapidly. Soon most of the small grains had been transformed into rocks—probably fist- to boulder-sized. But getting from boulders to a planet like the Earth is problematic: colliding boulders are more likely to fragment than to stick together, and gas remaining in the disk creates a drag that slows down boulders in their orbits and causes some to spiral in toward the Sun—ending up not as a planet, but as part of the Sun. Recent work suggests that turbulence in the disk may have played an important role in the agglomeration of boulders into larger bodies by swirling together clumps of boulder-sized rocks without violent collisions. When the clumps became big enough, gravity took over and the boulders coalesced to form "planetesimals"—loosely coherent precursors of the planets, perhaps a hundred miles or so across.

The best estimates from current research indicate that it took about ten million years, or less, to go from a cloud of gas and dust in space to an early Sun surrounded by planetesimals. As the planetesimals continued to grow by attracting more material from their surroundings, the largest of them out-competed their neighbors and swept up everything in their vicinity to become planets. The constant rain of rocks, boulders, and larger planetesimals onto the surfaces of the growing planets heated them up rapidly, and the outer parts of some of them may have completely melted.

Amidst this chaotic and violent process, the Earth, which would become the largest of the inner terrestrial planets, was growing quickly. It too was heating up as impacting bodies deposited large amounts of energy onto it. Our planet became so hot, in fact, that iron metal in the accreted material (recall that the chondrites contain abundant grains of iron) began to melt. Being very dense, the liquid metal sank to form the Earth's iron core. A raft of geochemical evidence, much of it uncovered over the past few decades and too extensive to examine in detail here, confirms that the metal core formed at this very early stage of the Earth's formation, as the planet was still growing. Today, 4.5 billion years later, the iron core is still partly molten, a relic of its early high-temperature period. The other terrestrial planets went through a similar process, segregating out iron cores, as did some planetesimals that ended up as asteroids rather than large planets. Iron meteorites like Cape York are probably fragments of such cores, launched toward Earth when their parent asteroids were broken up during violent collisions. These iron meteorites give us a hint about what our own planet's core may be like.

But even as the Earth reached almost its present size and had already segregated out iron to form a metal core, there was still an additional important chapter in its formation to come. Surprisingly, knowledge of this event came about through questions that earth and space scientists asked about the Moon—specifically, Why is the density of our close neighbor so different from the Earth's? And why, if the Moon was just a stray planetesimal captured into orbit around the Earth (this was once a popular idea for its formation), did analysis of rocks brought back by the Apollo astronauts show that the Moon has very close geochemical similarities to the Earth? These and other questions about the Moon's origin have been answered through an idea known as the giant impact hypothesis. It was first proposed in the 1970s as analytical data from the first returned lunar samples were published, but since then it has been refined and strengthened by many lines of evidence.

At the heart of the giant impact hypothesis is the idea that toward the end of planet formation in the solar system, a large leftover body nearly the size of Mars smashed into the Earth. Not only did the collision almost destroy the young Earth, it also blasted material out into space, material that was partly or mostly vaporized by the energy of the impact but eventually cooled and coalesced to form the Moon. The planet-sized impactor has even been given a name: Theia, after the Greek deity who gave birth to the moon goddess.

The giant impact hypothesis is currently the most plausible theory for the origin of the Moon. Evidence from isotope studies carried out on both Earth and Moon rocks indicates that the collision must have occurred between about forty and sixty million years after the solar system was born. By that time much of the Earth's iron had already settled into the core, and most likely the same process had occurred within the impacting body. For that reason, the material blasted out into space would have come only from the outer, rocky parts of both these bodies: simulations of the collision indicate that the impactor's dense metallic core would have traveled right through the outer part of the Earth and merged with our planet's core. Such a scenario is entirely consistent with the low iron content, and consequent low density, of the Moon.

The Moon-forming impact added more material to the Earth than it blasted away, effectively completing the heavy construction phase of building our planet and bringing it up very close to its present size. Bombardment with small and large objects continued—as we saw earlier, even now about a hundred tons of extraterrestrial material accumulate on the Earth every day—but significant growth had largely come to an end. What did this early Earth look like? We don't really know because we have no surface rocks from that time period. But lunar rocks have provided some clues.

One of the early surprises to come out of studies of lunar rocks was evidence that much of the outer part of the Moon had been molten very early in its history. This was a novel concept for planetary scientists, who refer to this molten outer layer as a "magma ocean" (*magma* is the geological term for molten rock). Inferring the presence of an early lunar magma ocean is possible only because the Moon is geologically inactive, which means that many of its earliest-formed rocks are still preserved. Work on samples collected during the Apollo missions shows that the highland areas—the light-colored patches on the Moon's face—are remnants of a rocky crust that once encircled the Moon and that formed as the magma ocean cooled and crystallized. Once the concept of a magma ocean had been articulated, earth scientists began to ask whether a similar phenomenon had occurred on Earth. Particularly immediately after the Moon-forming impact, our planet may well have been hot enough to sustain a surface magma ocean, although no unequivocal evidence has yet been found to prove its existence.

Whether or not there was a magma ocean on Earth, the process of chemical differentiation—the same process that caused iron metal to melt and sink to the interior, pushing rocky, relatively iron-poor material outward to form a surrounding mantle—also resulted in formation of a thin outer crust on our planet. But the chemical and mineral makeup of this crust is quite different from that of the underlying material. A cutaway view of the Earth (figure 3) shows the three chemically distinct layers: core, mantle, and crust. Formation of the core and mantle, as described above, was fairly straightforward and easy enough to envision. The crust, however, is a different story. It makes up less than 1 percent of the Earth's volume, and, because it is less dense than other parts of the planet, it constitutes an even smaller fraction of the Earth's mass. If the Earth were the size of a large apple, the crust would be considerably thinner than the apple skin. But in spite of its small volume, the crust is an extremely important part of the Earth for us, because it harbors all the resources necessary for human civilization.

Why is the crust so different from the underlying mantle? That is a complex question, and the answer has only been found through decades of observation and experiment. A simple, although not entirely accurate, way to think about the crust-forming process is that it is a kind of distillation phenomenon operating on a large batch of starting material (the mantle) and producing a small, concentrated sample (the crust) made up of just a few of the starting components. Melting of the mantle minerals, and migration of the resulting, less-dense, liquid magma upward toward the surface, is the process by which this is accomplished. Among the things winnowed out of the mantle and transferred into the crust in this way are many of the key materials important for modern life, like aluminum and many rare metals. Water and various gases contained

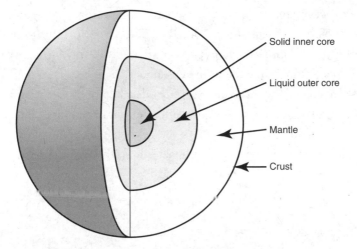

Figure 3. The interior structure of the Earth. The metallic core is composed mostly of iron, although it contains other elements as well, notably nickel. The core is slowly solidifying from the inside out; it has both an inner solid part and an outer liquid portion. The mantle, composed of rocky material, overlies the core and makes up most of the rest of the planet. Forming a thin skin at the Earth's surface is the crust, which is also made of rocky material, but with a very different mineral makeup and chemical composition.

in the minerals of the mantle are also extracted during this process, something that is strikingly illustrated by the volcanic rock pumice—its Swiss cheese texture is the frozen imprint of gas (mostly water vapor and carbon dioxide) that was dissolved in the magma but bubbled out as the liquid rock neared the surface. So in addition to forming the familiar rocks of the Earth's surface, the processes that made the crust have also been largely responsible for bringing the water of the Earth's oceans and the gases of its atmosphere to the surface.

In this brief chapter, we have traced what we know about the trajectory that began in an interstellar cloud and ended with our extraordinary blue planet, with its iron core, rocky mantle and crust, its deep, watery ocean, and its thick atmosphere. In cosmic terms, the process didn't take very long, and the Earth was essentially fully formed by roughly 4.5 billion years ago. It has continued to evolve in many ways since then, as we will explore in subsequent chapters. Over its lifetime our planet has also continued to attract bits and pieces of space material that remained in orbit around the Sun after the initial rapid interval of planet formation. As we will see in the next chapter, some of those bits and pieces have had a profound impact—literally—on the Earth.

CHAPTER THREE

Close Encounters

Even though it has been known for centuries that meteorites—some of them quite large, like the Cape York meteorite mentioned in the previous chapter—periodically crash to Earth from space, the possibility that impact cratering is a potent geological force, affecting the physical, chemical, and biological evolution of our planet, has only been appreciated relatively recently. Earlier, most scientists who thought about the problem at all realized that the Earth must have been bombarded by space debris in the violent early days of its existence, but assumed that the significance of the process diminished rapidly after this initial onslaught. Even the availability of ever-clearer telescopic images of the pockmarked faces of the Moon and Mercury did little to change this belief. Most of those craters, it was thought, were volcanic. This opinion began to change in the run-up to the Apollo program as the Moon came under more detailed scrutiny, but many earth scientists still did not understand the importance of impacts on the Earth. A primary reason was that there are very few terrestrial craters formed by impact, and controversy about their origins swirled around even these rare examples.

Near the end of the nineteenth century, one of the most interesting debates about the origin of a crater pitted a doyen of American geologists against a bull-headed businessman—and in the end, it turned out, the geologist was on the wrong side of the argument. The business-man was Daniel Moreau Barringer, a man who started his career as a lawyer but then decided to study geology, believing this would allow him to indulge his twin passions for the outdoors and the American West. Evidently he learned his geology well, because before long he was rich, the owner of several successful mining ventures. But Barringer was always on the lookout for new opportunities, and when he heard from a friend about a craterlike depression in western Arizona his ears pricked up. Many locals, his friend said, believed the crater had been formed by a giant iron meteorite because there were abundant small fragments of iron metal in and around the crater. Barringer knew that iron meteorites are virtually pure iron and nickel, much richer in these elements than any ore formed on Earth. If a huge iron meteorite lay buried beneath the three-quarter-mile-wide crater, it would be very valuable.

The problem was that one of the most respected scientists of the day, a man named G.K. Gilbert, the chief geologist of the U.S. Geological Survey, had already examined the crater and declared that it had been produced by a "steam explosion," not a meteorite. Like Barringer, Gilbert had heard about the iron fragments in the vicinity, and he initially considered that the crater might have been produced by impact. But tests he conducted during an expedition to the site in 1891 convinced him otherwise.

Neither Gilbert nor Barringer (nor anyone else at the time) knew much about the physics of large impacts, and this ignorance led both men astray, although in different ways. Their mistake was to assume that the impacting body would lie buried beneath the crater; in reality, it had largely vaporized during the collision. During his 1891 expedition Gilbert made measurements designed to detect the magnetic signal of a large, buried iron meteorite. The crater was large enough that he expected the signal to be quite strong, and when he found no magnetic effects at all he was confident he could rule out impact. The negative result caused him some difficulties, because he had to invent ad hoc and somewhat convoluted explanations for the presence of meteoritic iron around the crater (just coincidental, he said, the remains of an earlier meteorite fall) and for the complete absence of volcanic rocks at the site, which seemed to rule out a volcanic origin. His solution was to propose rather vaguely that hidden, "deep-seated" volcanic heat had caused a gigantic steam explosion that had excavated the crater.

Barringer was aware of Gilbert's conclusions, but he was a more intuitive investigator. He didn't have to conduct any experiments; when he visited the crater (he later claimed), he *knew* within a few hours just from its physical appearance that it had been formed by impact. So in the early 1900s, together with a partner, Barringer formed a company to mine the millions of tons of iron he believed were buried there. His early surveys and drill cores didn't locate a buried meteorite, but they did reveal that the local rocks had been crushed under massive pressure and thrown outward as an overturned blanket of ejecta, supporting the impact theory. Undeterred by his initial lack of success, Barringer redoubled his efforts to raise money for more exploration. In 1906 he and his partner each published a paper in the *Proceedings of the Academy of Natural Sciences of Philadelphia* outlining the results of their exploratory work at the crater. There was now no doubt, they concluded, that the crater had been produced by meteorite impact.

For the next two decades Barringer promoted his ideas to anyone who would listen, and simultaneously spent most of his fortune—and the funds of many wealthy investors—attempting to find the nonexistent buried meteorite. His partner, discouraged by the lack of results, eventually withdrew from the operation, but Barringer ploughed ahead. The geological data he gathered as he probed the crater convinced many in the scientific community that he was right about the meteorite impact, but for some reason, G. K. Gilbert and his colleagues at the U.S. Geological Survey refused to be drawn into the argument. They never commented publicly on the controversy, but a vocal minority of geologists still clung to Gilbert's steam explosion hypothesis. Inexplicably, a

1928 *National Geographic Magazine* article titled "The Mysterious Tomb of a Giant Meteorite" attributed the impact theory to Gilbert and, although it described the mining efforts in detail, didn't even mention Barringer.

Also in 1928—at the request of investors in Barringer's mining company—an astronomer named Forest Moulton did a detailed analysis of the crater and concluded both that the meteorite must have been much smaller than Barringer claimed and that it had probably mostly vaporized when it collided with the Earth. These conclusions sealed the fate of the mining venture, and in September 1929—although Barringer was in denial and resisted angrily—the directors of the company closed down activities. A few months later Barringer died of a heart attack.

In spite of the failure of his commercial venture, Barringer must be credited with recognizing and bringing attention to the first crater on Earth widely acknowledged as being of impact origin. In some ways, he has had the last word: the feature that consumed so much of his life, long known as Meteor Crater, is now officially named Barringer Crater. It is also still owned by the Barringer family, and is the site of a small museum dedicated to meteorites, impacts, and the history of the crater itself. If you happen to be in Arizona with some time on your hands, it is worth a visit.

Seen from the air, Barringer Crater is spectacular (figure 4). In the more casual days of air travel, pilots flying nearby would sometimes make a minor diversion from their flight plan to pass over it and bank the aircraft to give passengers a better view. The crater's remarkable preservation is due to its relative youth (age determinations show it to be just 49,000 years old) and the limited erosion in Arizona's dry climate. Researchers who have studied the crater estimate that the impactor initially weighed about 300,000 tons and had a diameter of almost 150 feet, but that about half its mass vaporized in the atmosphere before it even reached the ground. Most of the rest vaporized on impact. The crater, much bigger across than the impactor, was excavated by the shock wave surrounding the incoming material, not the meteorite itself.

Figure 4. Barringer Crater, Arizona, as seen from the air. The crater has a diameter of about three-quarters of a mile. (Courtesy U.S. Geological Survey; photo by D. Roddy.)

In one of those ironic twists of history, a geologist who studied Barringer Crater for his 1960 PhD thesis—and who permanently laid to rest the few lingering doubts that some had about its origin by impact—was the first recipient of the Geological Society of America's G. K. Gilbert Award, in 1983 (in fairness to Gilbert, it should be said that although he denied the impact origin of Barringer Crater, he was a proponent of impact as the source of most lunar craters). The geologist was Gene Shoemaker, and he is credited with almost single-handedly founding the field of astrogeology. Even before his work on Barringer Crater, Shoemaker had been interested in the Moon and the origin of its craters, and he went on to play a vital role in training Apollo astronauts for their missions. He also became a familiar figure to American TV viewers when, alongside Walter Cronkite on *CBS News,* he provided geological commentary on the Apollo moonwalks.

Though geologically recent, the Barringer Crater impact occurred long before recorded human history, and we have little direct information about its environmental effects beyond what can be discerned at the site today. But the destructive force of even small bodies from space is known from a much more recent collision: the so-called Tunguska event of June 1908, when an object estimated to be about 120 feet in diameter hurtled into the atmosphere above Siberia and exploded. No samples unequivocally identified with the original object have been found, so its makeup is unknown (some scientists have suggested it was a small, icy comet), and no crater was excavated. But the shock waves it generated completely flattened the dense Siberian forest over an area of nearly 800 square miles (figure 5). Fortunately, the Tunguska object struck an unpopulated area. Had it exploded over a major city, the disaster would have dwarfed recent tragedies like Hurricane Katrina, 9/11, or the 2010 Haitian earthquake.

A word about nomenclature is in order here. By definition, meteorites are space rocks that reach the ground; for that reason, I've referred to the rock (if that's what it was) that exploded above Tunguska simply as an "object," not a meteorite. The Barringer Crater impactor qualifies as a meteorite because pieces of the original object survived transit through the atmosphere. Small bits of space debris that completely burn up in the atmosphere (e.g., "shooting stars") are defined as meteors, but this is probably not an appropriate label for a larger object like the one at Tunguska. In what follows, I'll continue to refer to the Tunguska body either as an "object" or as a "meteoroid"—the latter a loosely defined term describing modest-sized bodies that move about in interplanetary space, whether or not they survive passage through the atmosphere when (or if) they collide with the Earth. Thus a meteoroid can become a meteorite—or a meteor. Confused? Scientists love to put things in categories.

At any rate, the object that caused the Tunguska blast was small by cosmic standards, but its effects were far-reaching. Atmospheric shock waves were detected as far away as Britain. Fine dust from the explosion spread through the atmosphere, reflecting light from the Sun (which in northern Europe doesn't dip far below the horizon at night in summer) and reportedly making it possible to read a newspaper outdoors at midnight in London. Perhaps because of its remoteness and the lack of detailed information about the event, the Tunguska explosion has

Figure 5. The site of the Tunguska meteoroid impact of 1908 in Siberia, photographed during a 1927 Soviet Academy of Sciences expedition led by the Russian mineralogist Leonid Kulik. The scorched and flattened trees, still very obvious almost twenty years after the event, attest to the impact's effects.

generated a raft of wild theories: it was caused by a UFO crash, by an errant black hole that collided with the Earth, by the detonation of a "natural hydrogen bomb" in an impacting comet, or by the explosion of vast quantities of natural gas that escaped from the Earth.

When it happened, however, news of the explosion trickled out slowly. Not until 1927, nearly twenty years after the event, did a scientific expedition, led by the Russian mineralogist Leonid Kulik, visit the Tunguska site. (Like Barringer in Arizona, Kulik thought he might find a large amount of iron metal at the site, and he used this possibility to persuade the government to fund his expedition.) The scientists had to slog through Siberian forests and late-spring snowdrifts, and they had to deal with superstitious local guides who feared visiting the devastated impact site. But eventually they did reach it, and made a detailed and thorough report. In addition to investigating the physical effects of the blast, Kulik's team gathered eyewitness accounts of the event from villagers who described a fireball as bright as the Sun streaking through the sky, ground-shaking explosions, blasts of hot wind, and smoke from smoldering vegetation. Some told of being knocked off their feet. One nearby village had a pragmatic response: they sent a delegation to a local priest, enquiring if such a never-before-experienced phenomenon was a sign that the end of the world was at hand. If so, they asked, what should they do to prepare? There is no record of the priest's response.

If the small Tunguska meteoroid could cause widespread local dev-astation, a large collision would be catastrophic. Although impacts are unlikely to spell the end of the world as we know it in the immediate future, they are—statistically—among the most dangerous of all natu-ral hazards because of their potential to affect a very large number of people. Fortunately, astronomers will probably be able to detect a large asteroid heading toward the Earth far in advance of its arrival. The question is, though, what can be done then?

As we will see later, there are many ideas about this, some promising and others less so. But before considering future mitigation strategies, it is instructive to turn to the rocks for clues about how impacts have affected the Earth in the past. The geological record, it turns out, has much to say on this subject.

According to a listing maintained by the University of New Bruns-wick in Canada, there are 176 impact structures on Earth. (Meteorite researchers refer to them as "structures" rather than "craters," because some, especially older ones, are heavily eroded and degraded, and no longer have the pristine appearance of a Barringer Crater.) Most of the 176 have been mapped and sampled, and some have been drilled to obtain information about their internal structures. Rigorous criteria are used to determine whether a particular crater is placed on the list. These go well beyond simply having a circular shape. The key evidence includes features produced by the high pressures and temperatures that are generated during a collision, such as crushed and broken rock, sometimes partly melted; minerals that result only from the passage of high-pressure shock waves; and peculiar features called "shatter cones" in the rocks of the surrounding countryside, diagnostic of shock waves spreading outward from an impact. Sometimes the presence of meteorite fragments confirms an impact origin, but at most craters there is little remaining evidence of the impacting body beyond minor chemical traces. As at Barringer Crater, the colliding objects are largely vaporized during the impact.

The youngest crater on the list is a small one produced by a meteorite that fell in Russia in 1947—it broke up into fragments in the atmosphere and fell as many separate pieces, one of which made a crater some eighty feet across and twenty feet deep. But all of the really large impact structures are old; the oldest known, which is also in Russia, is about ten and a half miles across and dates to approximately 2.4 billion years ago. Two very large structures, one in Canada and one in South Africa, have ages near 2 billion years. Since their formation, they have been substantially modified by geological processes, but their original diameters are estimated to have been in the range of 170 to 200 miles. Nothing larger has been found, although in 2006 researchers examining the gravity field over Antarctica discovered a large circular feature that *may* be an impact crater. It lies beneath glacial ice and has not been sampled, so it cannot be confirmed, but if it does turn out to have been produced by impact, it will be the largest such structure on the planet— about 300 miles in diameter, big enough to fit a state like Michigan or Ohio within its boundaries.

These large and very rare reminders of ancient impacts are little known outside the earth science community. However, there is one impact that has gained much wider attention: the one that "killed off the dinosaurs." The idea was first proposed in 1980, and the combination of a catastrophic, science-fiction-like impact and the extinction of the world's largest reptiles guaranteed widespread media interest. The story of how evidence supporting the impact theory was uncovered has been told many times before and I won't repeat it here, except to say that the conclusion is very firmly based in the discovery of a worldwide layer of the rare metal iridium in sedimentary rocks marking the boundary between the Mesozoic and Cenozoic eras. This boundary corresponds to one of the Earth's great mass extinction events, when not only dinosaurs, but also many other plants and animals died out. Conventionally—and somewhat confusingly for newcomers to the topic—the boundary is referred to as the "K-T" boundary, after the initials for the Cretaceous (spelled with a "K" in German) and Tertiary periods. (Compounding the confusion, the name *Tertiary* has recently been abandoned as a period in the geological timescale; it is simplest just to remember that the K-T boundary marks the end of the Cretaceous Period.)

Detection of the iridium-rich layer quickly led to the impact hypothesis because the metal is exceedingly scarce in the Earth's crust but greatly enriched in meteorites, and the only reasonable explanation for its high abundance in K-T boundary rocks is that it originated from a large extraterrestrial object. As pointed out by its discoverers—a group of researchers from the University of California at Berkeley, led by the geologist Walter Alvarez and his father, the Nobel Prize–winning physicist Luis Alvarez—iridium from a large, vaporized asteroid would have been ejected high into the atmosphere and dispersed globally. The Berkeley researchers argued that the simultaneity of the impact and the mass extinction could not be coincidence, and that the impact must be implicated in the extinctions.

The Alvarez team calculated the size of the impacting body from the amount of iridium at the K-T boundary, concluding that it must have been six to seven miles in diameter. Subsequent work has not significantly changed that estimate. One of the initial objections to the theory was that none of the Earth's known craters had the right age to be a candidate, yet a body of that size would certainly have left a large crater. This problem persisted for a decade after the initial discovery, but a great flurry of interdisciplinary research stimulated by the impact hypothesis led inexorably to the smoking gun, a crater 120 miles in diameter in the Yucatán Peninsula of Mexico. In spite of its size, this large feature had been completely missed because it is buried under more than two thousand feet of sedimentary rocks. There is no indication at the surface that a massive crater lies below.

Actually, the crater had not been *completely* missed. It was discovered in the 1940s by Mexican petroleum geologists using remote sensing techniques to map out subsurface geological features, but at the time they assumed it was a volcanic crater. Its existence was not widely known in the decades that followed, even among geologists. However, after the Alvarez group proposed their impact theory, earth scientists began to re-examine K-T boundary rocks for corroborating—or contradictory—evidence, and they soon discovered that sedimentary rocks from the Caribbean region contained more dramatic signs of impact than could be found anywhere else in the world, including layers of shocked and crushed rock presumably ejected from the crater, and abundant glassy spherules that appeared to be frozen droplets of the impact melt. Other sediments in the region showed evidence of disruption by huge waves, probably from tsunamis generated by the impact. Impact indicators were much less obvious at sites distant from the Caribbean, and the consensus was that the crater must be somewhere in the Gulf of Mexico region.

Then Alan Hildebrand, a geologist at the University of Arizona in Tucson, and a group of his colleagues reexamined the remote sensing data from the Yucatán structure and quickly realized that it was a very large impact crater. Importantly, borehole samples (like the remote sensing data, these had been collected during exploration for petroleum) showed that the rocks immediately overlying the structure were Paleocene in age—that is, they were deposited during the geological epoch immediately following the K-T boundary. In 1991 Hildebrand and his colleagues published their findings and proposed that the crater was the long-sought K-T collision site.

Since that time, a large amount of research has been done on the Yucatán crater, now known as Chicxulub Crater, after the name of a nearby town. That it was formed by impact is no longer in doubt, and many lines of evidence—including precise dating of its formation—link it unequivocally to the K-T boundary. The thick sequence of sedimentary rocks covering the crater makes it difficult to map out its features in detail, but the layers of overlying rock have also prevented erosion of the crater and preserved its original shape and features almost perfectly.

A map outline of Chicxulub Crater shows that it is almost exactly bisected by the coastline of the Yucatán Peninsula—half of it lies beneath land, the other half beneath the waters of the Gulf of Mexico (see figure 6). But at the time of the impact, the region was entirely submerged; the asteroid struck the shallow waters of the continental shelf off what is now Mexico, blasting through layers of water and ocean sediment before excavating the crater in the igneous rocks of the crust below. After an initial chaotic period—more on this below—more normal conditions returned and the crater began to fill up with layers of sediment. Later uplift of these sedimentary rocks has exposed them in the peaceful, tropical, present-day landscape of Yucatán. Hidden below that peaceful surface, however, is a tale of environmental disturbance of a magnitude only rarely experienced on Earth.

What happens when an asteroid or comet the size of the K-T impactor strikes the Earth? The phenomena affecting objects that enter the Earth's atmosphere from space have been examined rigorously in order to figure out how to bring things like the Space Shuttle, or an intercontinental ballistic missile, back to Earth safely without it burning up. The calculations have to be scaled up significantly for K-T–sized objects, but they nevertheless provide clues about what must have happened, and evidence stored in the geological record provides a way to cross-check the extrapolations. Extraterrestrial bodies strike the Earth at a variety of speeds and angles, depending on their trajectories in space, but all of them travel at many times the speed of sound—typically between about eight and twenty miles *per second*, or even faster. At such high velocities,

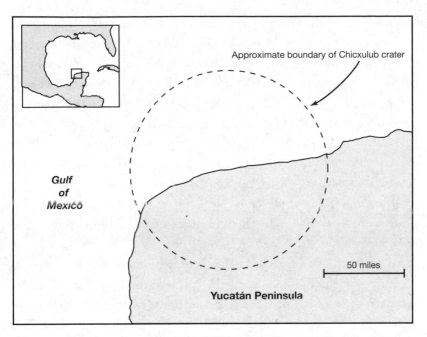

Figure 6. The location of the buried Chicxulub Crater in Yucatán, Mexico. Details about the crater are known from geophysical remote sensing data. (Based on data from Pilkington et al. 1994.)

the air in front of an incoming asteroid is rapidly compressed, exerting tremendous pressure on the object and heating it—or at least its surface—to incandescence (if you've ever inflated a bicycle tire manually with a small pump, you'll know that even modest compression quickly heats up air). The phenomenon is clearly recorded in meteorites, which not only lose much of their original mass through melting and ablation of their surfaces as they plunge through the atmosphere, but often retain a thin rind of melted material that geologists refer to as a "fusion crust."

Simulations of the K-T impact indicate that the asteroid shock-heated the atmosphere as it descended, and that the intensity of the heat radiated from the impact fireball itself was ferocious (with peak temperatures in the range of *tens of thousands* of degrees Fahrenheit at the core of the fireball). Close to the impact site, anything alive would have been incinerated instantly. Vegetation would have been set alight over a radius of perhaps two thousand miles—as far away from the Caribbean impact site as the present-day cities of Chicago, Montreal, San Diego, Lima, and Caracas. Furthermore, some of the material ejected from the crater was thrown upward at such high velocity that it broke out of the Earth's atmosphere into space, only to further heat the atmosphere as it reentered. The rain of ejecta fragments raised the temperature of the atmosphere globally, possibly desiccating forests and playing a part in the K-T mass extinction. Recent calculations suggest that for a short period—possibly a few hours—ground temperatures everywhere exceeded five hundred degrees Fahrenheit.

When they first suggested a link between the impact and the mass extinction at the K-T boundary, the Alvarez group proposed that the primary agent of extinction was the fine dust thrown into the atmosphere by the impact, which would have blocked sunlight and shut down photosynthesis for some unknown but extended period of time. They reasoned that widespread plant die-off would have severely disrupted the food chain, and that the effects would have been felt all the way up to large creatures like the dinosaurs. Darkness would also have caused sharply lowered temperatures, exacerbating the effects on all forms of life.

The dust scenario is plausible but difficult to prove; grains small enough to remain suspended in the atmosphere over long time periods are submicroscopic and very difficult to detect and quantify in the sedimentary rock record. But even if dust didn't completely darken the skies, there is an almost never-ending list of other environmental effects that would have made the Earth a very inhospitable place in the aftermath of the impact. Dense smoke from wildfires near the impact site and locally in areas of desiccated forests may have shut out the Sun instead. Rocks at the Chicxulub impact site are particularly sulfur-rich; when these were vaporized during the impact the sulfur was dispersed through the atmosphere, forming tiny aerosol particles in the stratosphere that further blocked sunlight (the dimming effect of such aerosols has been well documented from recent sulfur-rich volcanic eruptions, as we will see in a later chapter).

Some of the chemicals lofted into the atmosphere by the collision may have destroyed the protective ozone layer that encircles the planet, exposing life on the surface to deadly ultraviolet radiation for a short period. And for years after the impact, precipitation worldwide may have been strongly acid—a potent mix of sulfuric and nitric acid that some have compared to battery acid. (Nitric acid would have been produced from nitrogen oxides formed when the incoming asteroid shock-heated the atmosphere, and sulfuric acid from the vaporized sulfur-rich rocks of the impact site mentioned above. Both nitrogen and sulfur oxides dissolve readily in raindrops, producing acid rain.)

Locally, toxic compounds released by burning forests would have proved deadly to animals. Near the impact site, the shock wave caused an outward-spreading pulse of high pressure and blasts of wind many times stronger than those of the most extreme hurricanes, knocking over trees and stripping away soil. Because the impact occurred in water, it generated huge tsunami waves. In the Gulf of Mexico the largest are estimated to have been five or six *hundred* feet high, equivalent to the length of several football fields stacked end to end, and they washed more than a hundred miles inland. Their effects can be seen today in the sedimentary rocks of the region.

Finally, there were greenhouse gases. Layers of carbon-rich deposits like limestone were vaporized at the impact site, producing carbon dioxide. Extensive impact-induced wildfires may also have contributed significant amounts of carbon dioxide, and large quantities of methane, to the atmosphere. Combined, these additions probably approximately doubled the warming caused by greenhouse gases. Temperatures would have been on a roller-coaster ride: intense but very short-lived and locally variable heating due to the passage of the impactor and its ejecta through the atmosphere, abrupt cooling for a period of months or longer due to dust, smoke, and aerosols from the impact, then a prolonged period of warmth caused by the additional greenhouse gases. It is not surprising that many species succumbed to the effects of the K-T impact.

Remarkably, researchers now believe they have traced the object that caused the K-T catastrophe back to its source. If they are right, a collision in the asteroid belt nearly 100 million years before the K-T impact sealed the fate of the dinosaurs. The asteroid belt, which lies between Mars and Jupiter, is filled with millions of small and large rocky objects orbiting the Sun (the largest, Ceres, is almost six hundred miles across), and astronomers have recognized for some time that the belt is the major source of the Earth's meteorites. Occasional collisions, and gravitational perturbations of asteroid orbits by other solar system bodies, send these objects hurtling toward the Earth. Close similarities in mineralogical composition between asteroids and various meteorite groups confirm the connection. (The similarities are inferred using a technique called "reflectance spectroscopy," which involves analysis of the light reflected from an object. Because minerals absorb light of specific, characteristic wavelengths and reflect the rest, the peaks and valleys in the spectrum of light reflected from an asteroid serve as fingerprints for the minerals present on its surface.)

The largest rocky survivor of the asteroid belt collision thought to be the precursor to the K-T impact is known to astronomers as the asteroid Baptistina, which was discovered in 1890 by the French astronomer Auguste Charlois. Baptistina is about twenty-six miles across, but it was once much larger. It is, so to speak, the matriarch of an extended asteroid family, a large group of fragments all traveling together in roughly the same orbital path and, as far as can be determined from spectral analyses, all with a similar mineral makeup.

What is the evidence that connects the Baptistina family of asteroids to the K-T impact? It is, admittedly, circumstantial, but it is compelling. The idea was proposed in a 2007 report by a group of astronomers from the Southwest Research Institute in Boulder, Colorado, and Charles University in Prague, who used computer simulations to examine how the Baptistina asteroid family (BAF for short) had evolved over time.

They began by carefully mapping out the exact locations and sizes of all known objects in the family, and then, working backward, calculated how their positions have changed through time due to the gravitational pull of other bodies, including other asteroids, the Sun, and the planets. They found that the BAF's present-day configuration is best explained by breakup of a single large object (at least one hundred miles across) during an asteroid belt collision about 160 million years ago.

The calculations also showed that many fragments from the collision would have escaped the asteroid belt altogether, and some would have ended up in the vicinity of the Earth. Because of this, the researchers concluded, our planet must have experienced an increase in impacts for a period of about 100 million years after the collision. That is exactly what studies of both lunar and terrestrial craters had already shown: over that time period, the rate of impacts *did* increase, by a factor of two to four. The K-T collision took place during this period. While this does not prove that the impactor came from the Baptistina family, there is also another piece of corroborating evidence that bolsters the argument.

Spectral analysis of BAF asteroids indicates that they are composed of an uncommon type of extraterrestrial material that is especially rich in carbon; meteorites matching this composition are relatively rare in the world's collections. However, the fallout layer at the K-T boundary contains a unique chemical marker that is specific to this carbon-rich meteorite family. Thus both timing and composition fit. The conclusion that an asteroid "shower" generated by the breakup of Baptistina was the source of the K-T impactor seems highly likely; the astronomers who did the work estimated the probability at greater than 90 percent. They also concluded that around 20 percent of the large asteroids that are currently in the vicinity of the Earth originate from the BAF.

This story raises an important issue beyond identifying the source of the putative K-T dinosaur slayer: How likely is it that other asteroids will be diverted toward the Earth and collide, with equally devastating effects? In order to assess such threats accurately, the collision frequency for objects of various sizes must be determined. This can be done through astronomical surveys of objects in the Earth's vicinity (usually referred to as Near Earth Objects, or NEOs), combined with details of the past cratering record on the Earth and Moon. Fortunately, such data make it clear that the probability of a K-T–like impact in the near future is effectively zero, which should be a great relief to all of us. However, such analyses generally assume that the impact rate is approximately constant, and if asteroid "showers" periodically pelt the Earth with increased numbers of space rocks, that would not be true. Is the Baptistina shower unique, or are there other intervals in the past when impact rates were much higher than the long-term average?

In fact, the geological record shows there *have* been other times when the Earth experienced increased rates of meteorite bombardment. One interval in particular stands out; it dates back to about 470 million years ago, during the Ordovician period. In some ways the discovery of this episode is even more amazing than the findings about the Baptistina asteroid family. The story begins in the mid-1990s, when Birger Schmitz from the University of Gothenburg in Sweden and several of his colleagues discovered numerous fossil meteorites in a Swedish limestone quarry. This was almost unheard of; meteorites are rare objects to begin with, and most people have never seen one except in a museum. The odds of finding even one meteorite trapped in ancient limestone are minuscule, but in their initial search Schmitz and his colleagues recovered about seventeen pounds of these fossil space rocks, each of them small (typically an inch to a few inches across) and badly corroded, but nevertheless clearly identifiable as a meteorite. Because the meteorites occur only within a limited interval of the quarry's limestone layers, Schmitz and his co-workers concluded that they must have fallen over a period of less than two million years. They called the quarry "one of the most meteorite dense areas known in the world."

Since the initial discovery, many more fossil meteorites have been unearthed from Swedish quarries, all from rocks dating to about 470 million years ago. Based on their chemical compositions, most of these objects belong to a common meteorite type known to researchers as

"L chondrites" (the "L" stands for "low iron"). An important characteristic of the L chondrites is the presence of extensive shock effects, which are thought to have been produced in a violent collision that broke apart their parent asteroid. Dating studies of many different L chondrites show that the collision occurred very close to 470 million years ago. This coincidence of timing and composition with the influx documented by the Swedish quarry meteorites suggests that there was a short interval immediately after the collision when the Earth was showered with L chondrites.

But is there any other supporting evidence that this really happened? The answer is an unequivocal yes. By using computer simulations to trace orbits of individual asteroid family members backward in time, just as was done for the Baptistina family, researchers have discovered that the so-called Gefion asteroid family was likely formed by the breakup of a single object approximately 500 million years ago. Spectral analyses show that the Gefion asteroids have compositions similar to the L chondrites, and the computer simulations indicate that the Earth would have been pelted with ten to a hundred times the normal number of meteorites shortly after the collision that produced the Gefion asteroids—and that some of these impacting objects were large enough to leave craters a mile or more in diameter. In spite of the poor preservation of craters of this age, about a dozen—all more than a mile across—have been dated to the interval between 450 and 500 million years ago. This is a large number for a fifty-million-year period, and it corroborates the conclusion from the Swedish quarries that the Earth experienced a substantially enhanced bombardment rate at that time.

A consensus view of the overall impact hazard, based on studies of craters and NEOs, is shown in figure 7. The data indicate that on average the Earth experiences a very large impact like the one that occurred at the K-T boundary only once every 150 million years or so. Smaller objects collide much more frequently, however. A Tunguska-sized event is likely to occur about every one thousand years, and an asteroid as big as a multistory house—traveling at supersonic speeds—is expected to

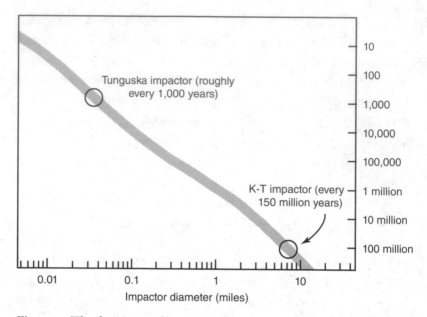

Figure 7. The frequency of impacts on Earth as a function of asteroid size, based on observations of Near Earth Objects (NEOs) and the terrestrial cratering record (after data in Chapman 2004). Note that both axes have logarithmic scales.

strike the Earth about once per century. Like the Tunguska meteoroid, such an object would probably explode before hitting the ground, but it would still cause substantial damage in a populated area. It is worth remembering that such predictions are based on average values, and although the probability is very low, it is entirely possible that a Tunguska-like impact could occur tomorrow, or next week—or not for several thousand years. Especially for rare events, statistics can be misleading.

Although quantifying risk and making predictions is important, in some ways what really matters is identifying specific hazards. That logic prompted the U.S. Congress in 1998 to direct NASA to catalog potential natural threats from space, and later, in 2005, to give the agency a specific mandate: to detect 90 percent of all NEOs with diameters of

140 meters (150 yards) or more by 2020. Both the British government and the United Nations have also set up programs to assess the hazard and investigate mitigation methods. By mid-2010 just over seven thousand NEOs had been cataloged (between four hundred and five hundred objects have been added to the total each year over the past few years). About eight hundred of these asteroids are larger than one kilometer (about two-thirds of a mile) in diameter and potentially civilization-destroying. However, none of these are currently forecast to impact the Earth. At present only one asteroid, estimated to be about 130 yards across, is in the "needs careful monitoring" category in NASA's "impact risk table," meaning that there is a possibility that it could come very close to the Earth within the next one hundred years—the timescale used in NEO surveys to identify potential threats.

The statistics for NEOs seem reassuring, but they are not grounds to be complacent. Telescopes can only observe minute portions of the sky at a time, and many asteroids—especially small but still potentially dangerous ones—remain undiscovered (I have only mentioned asteroids here; comets also pose impact threats, but of the seven-thousand-plus NEOs discovered so far, only eighty-four are comets). However, when a new NEO is discovered, the procedures for assessing its potential hazard are firmly established. Anyone who spots an NEO, including amateur astronomers, can report their findings to the Minor Planet Center (MPC) at the Smithsonian Astrophysical Observatory in Cambridge, Massachusetts. The center, which operates under the auspices of the International Astrophysical Union, verifies and regularly publishes information on identified NEOs. Separately, two different groups take the MPC data and calculate, via automated computer programs, the orbits of each reported NEO over the next century, and assess the possibility of impact. The Web sites of these groups are regularly updated, and if you are paranoid about being hit by an asteroid you can peruse them to get the latest information (the two prediction groups are the Near Earth Object Program at the Jet Propulsion Laboratory [JPL] in Pasadena, California, and the Near Earth Object Dynamics

Site [NEODyS], operated jointly by the universities of Pisa in Italy and Valladolid in Spain).

Just how effective these programs are was demonstrated in early October 2007 when an astronomer working near Tucson, Arizona, discovered a very small NEO (a few yards across) and reported it to the MPC. Initial calculations indicated that the object was on a collision course with the Earth, so the MPC immediately notified the astronomical community and NASA. Within an hour of receiving the initial data from the MPC, the Near Earth Object Program predicted that the asteroid would enter the Earth's atmosphere over Sudan early the following morning, only about twenty hours after it was first detected. NASA alerted various U.S. government agencies and issued a press release. The prediction proved accurate. The meteorite entered the Earth's atmosphere at the predicted time and exploded about twenty-five miles above Sudan. Satellites recorded the explosion, and it was also observed by a commercial airline pilot who had been alerted about the incoming object.

The Sudan meteorite was not a rare event; several objects of this size strike the Earth every year. What was unprecedented was that it was spotted before impact. As soon as word of its presence was broadcast, astronomers around the world raced to their telescopes and began making observations. Their data flooded into the MPC and was used to update the accuracy of the meteoroid's trajectory in real time. The precise location information also made it possible for Peter Jenniskens, an astronomer from California, to fly to Sudan after the impact and quickly find surviving pieces of the object on the desert floor. Using the tracking data from the MPC, he and a group of students from the University of Khartoum combed the region where the meteorite was calculated to have hit the ground, and found numerous fragments. It was the first time samples of a meteorite that had been observed in space had actually been picked up on the Earth's surface. Subsequent expeditions to the same area have brought the total number of recovered pieces up to several hundred.

Improved monitoring of NEOs means that they sometimes make the headlines. In March 2004, astronomers announced—and the press duly reported—that a "record-breaking" near miss was about to occur: an NEO roughly one hundred feet across would pass by the Earth at a distance of only 26,500 miles within a few days. That seems like a large distance, but on a cosmic scale it is not very far, only slightly more than the circumference of the Earth and much less than the distance to the Moon. But in reality near misses from objects of this size occur on a regular basis—at least once every few years. (On March 2, 2009, another object of about the same size whizzed by our planet at just under twice the distance of the 2004 NEO.) "Small" objects like these are typically only detected when they are very close to the Earth, if they are detected at all.

Later in 2004, however, a much more serious threat was reported. Based on NEO observations, there was, astronomers estimated, an almost 3 percent chance that a fairly large asteroid—measuring 700 to 1,100 feet across and thus many times bigger than the Tunguska object— would collide with the Earth in 2029. The story of the asteroid—known to astronomers as "99942 Apophis"—hit the headlines. But media attention quickly waned when additional, more detailed analyses of the asteroid's orbit showed that the probability of impact was actually much lower than originally calculated.

This story illustrates just how difficult it is to predict collisions with the Earth. Although we tend to think of our planet as a very large place, in reality it is a tiny target in what some have referred to as the "cosmic shooting gallery." Typically, large NEOs are spotted when they are tens of millions of miles away, and their motion is observed over a short time interval. An object's pathway through space has to be calculated far into the future from that very small recorded segment of its orbit. Even minute errors—in the measured path, the asteroid's size, how it rotates, the gravitational attraction of planets and other asteroids in its vicinity, or several other factors—can significantly alter the calculated future location of the object ten, twenty, or a hundred years from now. In addition, the subtle, long-term effects of the Sun's radiation can gradually alter the orbit of an asteroid in ways that are difficult to predict.

For Apophis, the initial telescopic observations indicated a significant possibility of collision on a very unlucky Friday the thirteenth (April 13, 2029). But subsequent observations used radar, which can track the path of an asteroid more accurately than optical observations, and they greatly reduced the original uncertainties in the orbit calculations. It is now possible to say with confidence that Apophis will not hit the Earth. However, it will be a close encounter. At its nearest, Apophis will be only 18,300 miles away, close enough that if you happen to be around on April 13, 2029, you may be able to see it zooming by even without a telescope.

As NEO detection capabilities outstrip our ability to track orbits accurately, there are likely to be more false alarms of this kind. But what if the new observations had confirmed that Apophis was truly on track to collide with the Earth? Could anything really be done? That problem has challenged a small group of scientists and engineers ever since the impact hazard became widely recognized. As far as is known, the very first full-scale engineering investigation of this question was done as a class project in 1967, by students at the Massachusetts Institute of Technology (MIT). They were given the task of preventing the real-life asteroid Icarus (which is about a mile across) from hitting the Earth if it were on a collision course (it isn't, but its orbit regularly brings it relatively close to the Sun—hence its name—and the Earth). The exercise became known as Project Icarus, and the students came up with a brute force solution: they proposed sending half a dozen rockets to the asteroid, each carrying a nuclear bomb, and detonating them.

Nuclear explosions still figure in the arsenal of tools that might be used to deflect an asteroid, but as more has been learned about these space objects, attention has turned to more subtle approaches. Faced with a potential impact, there are really just two choices: completely destroy the object, or divert it. If the lead time is long, diversion is the best option by a considerable margin. With ten years' warning, for example, an asteroid's speed would only need to be altered by about half an inch per second to engineer a miss instead of a collision. That change is only a tiny fraction of the velocity of most asteroids relative to the Earth, which ranges up to sixteen or seventeen *miles* per second.

Both diversion and destruction become increasingly difficult for larger asteroids, but diversion remains the preferred approach; the trajectories of pieces thrown out by an explosion can't be predicted with any accuracy, and simply blowing up a large asteroid could end up showering the Earth with hundreds of smaller but still dangerous fragments. Some imaginative proposals have been made for breaking up an asteroid into pieces small enough to pose no hazard even if they do reach the Earth, without blowing it apart, but they are not (yet) feasible solutions. They include such bizarre possibilities as gigantic "cookie cutters" that literally slice up an NEO into small chunks, or "eaters" that transform it into dust.

However, diverting an incoming asteroid by changing its orbital path clearly seems to be the most promising solution. Many ideas have been investigated, including explosions on the asteroid's surface or in space nearby that would impart a strong, short-lived impulsive force and alter the asteroid's trajectory but would not break it up. In principle, simply crashing a spaceship (or a series of spaceships) into an asteroid would accomplish the same thing, although it would work only for small objects. Even a large spaceship would have no effect on a K–T–sized asteroid—it would be the cosmic equivalent of a fly smashing into the windshield of a speeding automobile. For this reason, the use of slow, long-term force is a far more attractive solution. That might require anchoring some sort of propulsive device to the asteroid, or it could entail covering its surface with material that would either absorb or reflect sunlight to make use of solar energy as the driving force. Although appealing in their simplicity, these approaches are complicated by the fact that all known NEOs rotate rapidly. To push an asteroid in a particular direction, a fixed propulsion device would have to be turned on and off as the object rotated. The effects of surface coverings would have to be calculated very carefully based on accurate knowledge of the times during which different surfaces of the asteroid are in sunlight or in shadow.

When the problem of preventing an asteroid collision was initially addressed, it was assumed that incoming bodies would resemble the meteorites that reach the Earth's surface: hard rocky or metallic objects. As more was learned, this idea was reinforced by the seeming similarity between common meteorite and asteroid types. But while the mineral makeup of asteroids and meteorites may be the same, recent observations suggest that many asteroids are just loosely coherent "rubble piles," not single, solid, rocks. And quite a few have also turned out to be paired objects rather than single NEOs. Both features complicate the task of diverting a potentially hazardous asteroid. It has become increasingly clear that any such efforts will require detailed knowledge of the object's physical properties.

That realization has made the exploration of asteroids an important goal of space programs around the world. Already it has led to two separate successful landings on NEOs, both truly incredible accomplishments. The first attempt began in February 1996, when NASA launched a mission to the well-known asteroid Eros, which was discovered by European astronomers in 1898 and has been studied by ground-based observers ever since. Just how complex such projects are is illustrated by the unforgiving launch window available to the mission: the asteroid's orbit dictated that there were only twelve suitable launch days, each with only about *one minute* during which the launch could be made successfully. But after some nail-biting problems along the way, the spacecraft touched down safely on Eros on February 12, 2001.

Eros is a small, oblong object, just twenty-two miles in maximum dimension, that spins around its own axis once every five hours. For a full year before landing, the spacecraft orbited the asteroid, making measurements and taking spectacular images of its heavily cratered and boulder-strewn surface (figure 8). Eros is one of the solid, rocky varieties of asteroid, and touchdown of the spacecraft on its hard surface was gentle and picture-perfect. The mission scientists and engineers back

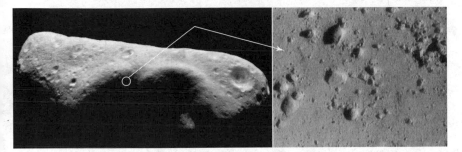

Figure 8. Two images of the asteroid Eros. The left panel is a mosaic of pictures taken by NASA's *NEAR Shoemaker* spacecraft as it orbited around the asteroid on November 30, 2000. The width of the asteroid as seen here is about twenty-one miles. On the right is one of the last images recorded as the spacecraft descended to the asteroid's surface on February 17, 2001, taken from an altitude of 820 feet. The field of view is about thirty-nine feet across. (Courtesy NASA/JPL-Caltech.)

on Earth were ecstatic: the landing was a bonus, because it had not been included in the original mission plan. The lander continued to send back signals from the asteroid surface for more than two weeks.

The second asteroid mission was launched in 2003 by the Japan Aerospace Exploration Agency. The spacecraft, called *Hayabusa* ("falcon" in Japanese) had an even more ambitious task than the Eros mission: to return asteroid samples to Earth. Although *Hayabusa* experienced various technical problems along the way, it successfully touched down (very briefly) on a small asteroid called Itokawa, less than half a mile in maximum dimension, not just once, but on two separate occasions six days apart, before beginning the long journey back to Earth. Late on the night of June 13, 2010, *Hayabusa* streaked across the sky over Australia, burning up as it entered the Earth's atmosphere but dropping its heat-shielded sample capsule safely into the desert. The next morning it was recovered, and soon it was back in Japan. A little over a week later scientists began unpacking the sample container, and in November 2010 they announced that by using an electron microscope they had found about 1,500 tiny particles in one of *Hayabusa*'s two sample capsules. Their analyses confirmed that most of these grains are indeed from Itokawa's surface. Returning samples of a tiny asteroid to Earth is one of the most impressive accomplishments so far in the annals of space exploration.

Hayabusa sent back a stream of data about Itokawa, including photographs showing it to be a lumpy, irregular object with giant boulders poking randomly out of its surface. It has a low density and is almost certainly a true "rubble pile" asteroid, composed of a loose agglomeration of fragments. This contrast between the properties of the only two NEOs that have ever been visited emphasizes the importance of knowing as much as possible about any asteroid that poses a potential impact hazard before attempting to deflect or destroy it: an Itokawa-like body would probably require quite a different approach from an Eros-like object. However, all of the methods so far proposed require spacecraft to visit the target, and that capability has now been demonstrated. Although the probability of a catastrophic impact in the immediate future is small, it is comforting to know that with sufficient warning and proper planning this particular geological hazard can probably be averted.

CHAPTER FOUR

The First Two Billion Years

Several important events from the Earth's history, such as the Moon-forming impact and the K-T mass extinction, have been discussed in earlier chapters, but here I'd like to begin a more systematic walk through our planet's geological past, interspersed in later chapters with more detailed discussions of phenomena such as earthquakes and climate change. Space constraints mean that only selected highlights of the Earth's history can be discussed. But I hope that this abbreviated treatment will provide a sense of our planet's fascinating history, and how that history can inform us about the ways in which Earth processes operate. I hope also that it will give readers an understanding of the enormity of the geological change that has affected the Earth over the past 4.5 billion years.

It may seem cavalier to devote just one short chapter to the first two billion years of the Earth's history, almost half the span of our planet's existence. Whole books could be filled with the knowledge earth scientists have accumulated about this time period, and yet we still know much less about it than we would like. The problem is not lack of effort, but simply that nearly all rocks formed during the first billion years, and even many of those that originated during the second billion, have either been completely destroyed or, at a minimum, highly altered.

Erosion has ground down and washed away ancient mountain ranges; collisions between tectonic plates have thrust surface rocks deep into the Earth, heating, folding, and metamorphosing them almost beyond recognition. In some cases, this has happened multiple times. The resulting gaps in the geological record greatly complicate the task of reading the rock record in the most ancient parts of the Earth's crust, but in spite of this much has been learned.

For geologists, discovering the Earth's oldest rocks is akin to finding the Holy Grail. Currently the record is held by a group of rocks exposed along the eastern shore of Hudson Bay in Canada. If you go there during the brief northern summer, and are willing to brave clouds of mosquitoes, you can sit on an outcrop of these gray, nondescript rocks and contemplate the surrounding silent landscape, barren but beautiful. The rocks themselves are far from silent, however; bit by bit they are yielding information about the long history they have experienced. They are part of a formation known to geologists as the Nuvvuagittuq Belt—a tongue-twisting name from the local Inuit language—and they date to 4.28 billion years ago. That means they formed in the midst of the Hadean eon, less than 300 million years after the Earth itself was born.

The age of the oldest known rocks has been gradually creeping upwards. When I was a student, 3.6 billion years was about as old as they got; holding a piece of the Earth's crust that ancient in your hand was a marvel (and it still is). I recall one internationally renowned professor whose specialty was Precambrian rocks telling us that it was unlikely that anything much older than 3.6 billion years would ever be found. There probably was older crust, he said, but it would be unrecognizable; it would have been reprocessed or melted by geological processes over the long span of the Earth's history. But as radiometric dating techniques improved, and as geologists in the field began to realize that within the swirled and deformed masses of ancient metamorphic rocks there might exist preserved remnants of even earlier crustal rocks, older ages began to pop up in the geological literature. First, fragments of metamorphosed ancient ocean floor from the Isua region of west Greenland yielded dates near 3.8 billion years. Then some bands of metamorphic gneiss from northwest Canada were dated at 3.9 billion years. More detailed studies of these same outcrops revealed that they were really a mixture of different rocks formed at different times, possibly smashed together in an ancient collision between tectonic plates. The oldest bits yet found in this mix are just over four billion years old. And then, in 2008, came the news about the 4.28-billion-year-old rocks from the eastern shore of Hudson Bay.

There is still some controversy about this record-breaking age. The scientists who analyzed these rocks took an unusual approach, employing a dating method not normally used for terrestrial rocks. Some earth scientists will not be convinced that these rocks are as old as claimed until there is corroborating evidence from another technique. But the data are clear about one thing: even if the Hudson Bay samples crystallized more recently than 4.28 billion years ago, the analytical results still show that they contain precursor material that *is* that old. Chemical analyses also show that the rocks, although since metamorphosed, were originally volcanic. So regardless of how the date for these rocks is interpreted, the evidence indicates that crust-forming volcanism occurred on the Earth at least 4.28 billion years ago.

That conclusion is confirmed by data from the opposite end of our planet, the scrubland of Western Australia. There the metamorphic Precambrian rocks are much younger than those found on the shore of Hudson Bay—they are "only" 3.6 billion years old. But among them are bands of quartzite, a metamorphic rock made up predominantly of the mineral quartz, and also containing rare small crystals of zircon, the mineral of choice for uranium-lead dating. The ultimate precursor of the quartzite was beach sand; the zircon crystals, like the quartz grains, are part of the winnowed residue of material weathered out of even older rocks and deposited as sand along an ancient shoreline. And because the zircon grains are especially robust, they retain the age of the precursor rocks. Extracting them is a painstaking task: large quantities of quartzite (a very hard rock) have to be crushed and sifted to recover even a small number of grains, and the crystals must be analyzed one by one. But the effort has been worth it. Several zircon crystals from the Australian rocks have been found to have ages greater than four billion years, and one has been dated to 4.4 billion years.

This very ancient date refers to one small crystal and not an expanse of rocks like those exposed in northern Canada. But the chemical characteristics of this single zircon grain indicate that it formed within a rock similar to the granite that today makes up much of the continental crust. That means crust-forming processes were active not long after the Earth itself formed—much earlier than many geologists had originally suspected. It also provides an additional incentive, if one is really needed, for geologists to keep searching for localities where larger, rocky fragments of the Earth's earliest crust might be preserved.

Just what was the Earth like during the first few hundred million years after its formation? We don't know for sure, but we can make some well-informed estimates. Perhaps surprisingly, some of the information bearing on the very earliest days of our planet's history comes not from the Earth itself, but from the Moon.

The gigantic collision that spewed masses of molten and vaporized rock out into space, creating a disk of hot material around the Earth that coalesced to form the Moon, was described in chapter 2. The collision also added enough material to the still-growing Earth to bring its mass up to about 99 percent of its present value. Although orbiting rocks and small planetesimals continued to bombard our planet, it would never again be hit by anything nearly as large as the Mars-sized body of the Moon-forming impact. Thus the date of the giant collision is a crucial number in the chronology of Earth formation. And in 2007, isotope analyses of lunar rocks placed the time of impact at 4.5 billion years ago. This date is consistent with independent data from the Earth, which indicate that our planet was nearing its present size and had segregated an iron core a few tens of millions of years before this.

As described in chapter 2, one of the earliest and most important

findings from the study of rocks brought back by the Apollo astronauts
was that the entire outer portion of the early Moon was molten, quite
literally an ocean of magma. Rocks from the Moon's ancient highland
areas, which are the remnants of the crust formed as the magma ocean
cooled and crystallized, have been extensively probed for any light they
might shed on the Moon's earliest history. Extracting clues from these
rocks has not been as straightforward as you might imagine, however,
for while the Moon has no processes that heat and metamorphose
crustal rocks as happens on the Earth, the highland areas have been
heavily cratered by meteorite bombardment. Most of the old rocks have
been shocked and battered, and some have even been melted. But careful chronological studies show that the magma ocean had a solid crust
by 4.46 billion years ago, only about fifty million years after the giant
impact. The oldest sample from the Earth, the 4.4-billion-year-old
zircon crystal, is only slightly younger, so it's clear that if the Earth
also had an early magma ocean, it too had developed a crust by this
time, or even earlier (see figure 9 for a time line). It is worth mentioning
here that although I have given these dates as absolute numbers, each
of them has an "uncertainty" attached to it, typically in the range of
ten to a few tens of millions of years. This is a reflection of the reality
that analytical measurements—even when performed using the most
sophisticated modern instruments—are always uncertain by a small
amount. The emphasis should be on the word "small." Although ten
million years may seem like a huge span of time, it is only a few tenths
of 1 percent of the age of these materials.

What is remarkable about research on the Earth's oldest solid—that
4.4-billion-year-old zircon crystal—is that it has been possible to measure not only its age, but also many aspects of its chemical makeup.
The crystal is no bigger than a grain of sand, but by using state-of-the-
art analytical instruments capable of analyzing an area much smaller
across than the diameter of a human hair, geochemists have been able
to ferret out chemical information that provides clues about conditions
on the Earth 4.4 billion years ago. Principal among the conclusions of

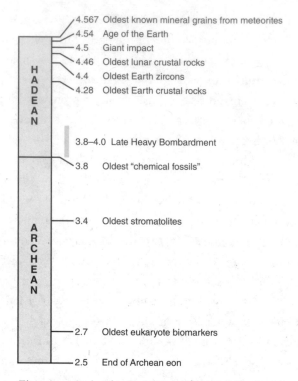

Figure 9. A simple time line of the Earth's first two billion years, starting with the oldest known objects in the solar system. Ages are given in billions of years; although they are accurate at the time of writing, some are subject to change as researchers uncover new information.

this work is that liquid water was present at the surface. Perhaps that doesn't sound too startling, given the ubiquity of surface water today. But remember, this was only 100 million years after the giant impact caused large-scale melting. It was near the beginning of the Hadean eon, which lasted until 3.8 billion years ago and was a period of frequent bombardment from space and probably of high temperatures at the surface. Before work on the ancient zircon crystals proved otherwise, many geologists thought that during this early part of the Earth's history any water would have existed only as vapor in a heavy, steamy atmosphere, not as liquid water in a primordial ocean.

And there is more. The most straightforward interpretation of the zircon grain's chemical characteristics indicates that the granitelike rock in which it formed was itself produced by the melting of even older rocks with quite complicated histories of their own; the precursor rocks had been weathered at the Earth's surface, buried, and then heated to their melting temperature. All this could have happened fairly quickly in geological terms; the original rocks may have been only a few million years older than those in which the zircon grew. We have no information about the chronology of events affecting the precursors, but their very existence means that some of the same crust-forming processes we know today (specifically, the remelting of older crustal materials to make granitelike rocks) were already in operation near the very beginning of our planet's history. Thus one tiny zircon crystal has opened a large window to the early Earth. The scene we can view through that window is quite different from the one geologists had imagined just a few short years ago.

What were the foundations of the apparently erroneous idea that the early Earth was too hot to sustain liquid water at its surface, perhaps all the way through the Hadean eon? Two factors were especially important. The first was the idea that a massive atmosphere, rich in water vapor because of the initially high temperatures, enveloped our planet like an insulating blanket, preventing cooling. The second was that continuous bombardment—the final phase of the accretion process—kept the surface of the Earth above the boiling point of water for many hundreds of millions of years.

There is no question that the Earth's formation left it very hot, and that the giant, Moon-forming collision melted parts of its outer regions and perhaps even the entire surface. The Earth is still cooling down from this fiery beginning, a fact we don't often appreciate from our vantage point on its surface. But other things being equal, even a magma ocean would start to congeal quite quickly in geological terms. Red-hot lava flowing down the sides of volcanoes such as Hawaii's Kilauea crusts over almost instantaneously and—depending on the thickness of the flows—can be stone cold (really) within a few days to perhaps a year or two. Compared to a Kilauea lava flow, of course, a magma ocean is vast, but its surface still would have solidified and cooled fairly rapidly.

Long after the Moon-forming impact, however, collisions of large and small bodies continued to deposit heat energy at the Earth's surface, although the process was sporadic and affected different parts of the Earth at different times. From studies of Moon rocks and dating of lunar craters, the bombardment appears to have tapered off rapidly beginning about 3.8 billion years ago. For about 200 million years before that—between approximately 4 and 3.8 billion years ago—the impact rate was very high, a phenomenon that has been labeled the "Late Heavy Bombardment," or LHB. Large, mountain-ringed basins on the Moon (like the easily visible Mare Imbrium) cluster in this age range. These basins are essentially giant holes punched through the Moon's original crust by impacting bodies, and they are now floored by basalt flows that welled up from the Moon's mantle. The objects that created these basins were many tens of miles in diameter, and an even heavier rain of similar bodies would have pummeled the Earth because of its larger size and therefore much greater gravitational attraction. Although the impacts would have melted the crust locally and evaporated the upper layers of the oceans globally, the evidence from the 4.4-billion-year-old zircon crystal indicates that just 150 million years or so after the Earth began to form, it maintained some amount of liquid water at its surface.

Because water is a primary ingredient for life, its very early presence immediately raises the question, When did life on Earth originate? The oldest true fossils are finely laminated structures known as stromatolites (see figure 10). These objects come in a variety of shapes and forms, from simple cones to large branching columns, and they are the predominant fossil type in all sedimentary rocks older than about 600 million years. The oldest stromatolites come from a formation in Western Australia with an age of just over 3.4 billion years.

Figure 10. *Left:* A close-up view of fine-scale laminations in an eroded, cone-shaped 3.43-billion-year-old stromatolite from Western Australia. The area shown is just under nine inches across. (Courtesy Abigail Allwood.) *Right:* Living, partly emergent stromatolite mounds growing in Shark Bay, Western Australia, March 2005. (Photo by Paul Harrison, reproduced here under the terms of the GNU Free Document License.)

Amazingly, stromatolites are still forming today, although only in a few localities. Close examination of these modern analogs of the ancient fossils has provided crucial insight into how and where they grow. The intricately layered structures are made up of thin mats of microbes—colonies of single-celled bacteria, including (at least in the modern examples) photosynthetic bacteria called cyanobacteria—that act as traps for sediment grains and gradually build up mounds or domes or columns. They grow in warm, shallow ocean water along the edges of continents, either completely submerged throughout their lives or partly emergent during low tides. By a quirk of fate, only a few hundred miles separate the sites of the world's oldest and youngest stromatolites: like the Archean examples, the living stromatolites are found in

Western Australia, at a place called Shark Bay (figure 10). Shark Bay was declared a UNESCO World Heritage site in 1991, and if you visit there you can view these strange, crenulated objects close up from a conveniently placed viewing platform.

At least seven different varieties of the 3.4-billion-year-old fossil stromatolites have been identified. The local distribution of these varieties and the types of rocks they are associated with indicate that they constituted a diverse ecosystem along an ancient shoreline that had recently been submerged due to a rising sea level, with different stromatolite types occupying slightly different environments. This paints a picture of an Earth on which life was already abundant by 3.4 billion years ago, and perhaps one on which—if the modern stromatolites are a reliable guide—bacteria were already consuming carbon dioxide and producing oxygen through photosynthesis. The abundance and diversity of the early stromatolites also suggest that life arose long before 3.4 billion years ago, possibly during the Hadean eon.

There are, however, no Hadean fossils; the only clues that life may have been present on the Earth before 3.4 billion years ago are indirect. The Earth's oldest known sedimentary rocks, which date to 3.8 billion years ago, have been found in both the Isua region of western Greenland and northern Quebec. These rocks have undergone multiple episodes of metamorphism, but they still retain features showing they were water-deposited, and although they contain no recognizable fossils the Greenland samples contain graphite, a form of pure carbon. Isotope analyses of the graphite show that it has a specific carbon isotope value that is characteristic of biological carbon, suggesting that it originated in living organisms. If the oldest known sedimentary rocks contain "chemical fossils" indicative of life, it is likely that our planet has been inhabited from very early in its history.

There are, however, some critics of the Greenland evidence. They do not question the isotope analyses, which clearly label the carbon as biological, but they point out that the graphite occurs in very small amounts, and that its precursor biological carbon could have been introduced long after the rocks formed, perhaps during one of the several metamorphic episodes that have affected them. But if we accept that the bits of graphite really did originate from organisms inhabiting the 3.8-billion-year-old ocean, that need not be a particularly startling conclusion. All the chemical elements important for life had been present on the Earth from its beginning, and there had been hundreds of millions of years for them to react in every imaginable way. Molecules and compounds formed and reformed in the primordial ocean, and when a few appeared that could reproduce themselves spontaneously, the evolution of life forms was almost inevitable.

The crucial step—the ability of a molecule to clone itself—has recently been demonstrated in laboratory experiments with artificially produced RNA (ribonucleic acid) molecules at the Scripps Research Institute in California. These engineered molecules can reproduce themselves very rapidly and almost endlessly. Even more interesting, different varieties of the RNA molecules "compete" with one another when they are put together in the same experiment; those that replicate fastest are the most "successful" in taking over the environment. Although the experimental molecules reproduce themselves, they cannot evolve and are not living organisms. But their behavior does provide a glimpse of the kinds of processes that may have been precursors to the rise of life on the early Earth.

The bacteria responsible for the earliest fossils, the stromatolites, were primitive in the sense that they were single-celled organisms without a defined nucleus and with few other internal structures. Together with another group of microbes, the so-called archaea, they were the only living things on the planet for most of the Earth's first two billion years. Both archaea and bacteria are still numerous. You need a microscope to see them, but there are so many of these organisms on the Earth today that if you totaled up their weight, you would find that they make up a large fraction of all living material.

The plants and animals we are familiar with are composed of cells that are more complex than those of the early single-celled organisms.

The main difference is that they contain a discrete nucleus within which many vital cell functions occur. All organisms made up of such cells—including us—are called "eukaryotes." The very first unambiguous fossil record of eukaryote cells comes from rocks of the Proterozoic eon, but it is probable that they evolved much earlier. Like the evidence for early life in the Greenland rocks, clues about the first eukaryotes come from chemical tracers, not physical fossils. Biological compounds specific to eukaryote cells have been found in sedimentary rocks dating to 2.7 billion years ago. Referred to as "biomarkers," these molecules are robust and not easily degraded, and unless they were somehow introduced into the sedimentary rocks much later, either naturally or as a result of contamination during handling (and the scientists who made this discovery were meticulous in assessing and ruling out this possibility), they are a clear signal that eukaryotes joined archaea and bacteria in the oceans toward the end of the Archean eon.

Inextricably tied up with the appearance and evolution of life on the early Earth is the question of what the oceans and atmosphere were like, and when and how they began to change toward their present states. The most significant difference between then and now is in oxygen content. There is very strong evidence that oxygen, which currently constitutes just over one-fifth of the volume of the atmosphere, was present only in vanishingly small amounts (probably less than one-tenth of 1 percent of the present-day levels) at least until the end of the Archean eon 2.5 billion years ago. That is nearly half of the Earth's history, and such a state of affairs has wide ramifications.

How do we know about oxygen levels on the early Earth? The evidence, as for so many other aspects of our planet's history, is stored in rocks. One clue is that certain kinds of sedimentary rock formations, called banded iron formations (BIFs), are common among Archean rocks but rare from more recent times. The world's oldest sedimentary rocks, the 3.8-billion-year-old examples from Quebec and western Greenland, contain this kind of deposit, but no BIFs are forming today—or have since the Precambrian. To understand what this has to do with oxygen in the atmosphere requires a short foray into some simple chemistry.

The first thing to note is that oxygen is very reactive, so much so that in an early oxygen-poor world, chemical processes both in the ocean and on land were very different from those that occur today, and those differences are reflected in the chemical characteristics of the ancient sedimentary rocks. Iron—one of the more abundant elements in the Earth's crust—presents a good example because its behavior depends very sensitively on the amount of oxygen in the environment. If oxygen is abundant, iron reacts rapidly. Try leaving something made of iron out in the backyard for a while, and you will see the process of oxidation in action: rust is just oxidized iron. In the ocean, dissolved oxygen reacts with dissolved iron, causing it to precipitate out as a rusty iron oxide coating on sediment grains. This process rapidly removes iron from today's oceans and insures that seawater contains very little dissolved iron. But the massive amounts of iron in the archean BIFs require that ancient seawater contained much higher amounts. That could only have happened if the atmosphere and oceans contained little or no oxygen.

BIFs are economically important—they are the primary source of iron ore—and because of this they have been studied in considerable detail. The ore is composed of iron oxides, so it is clear that oxygen, as well as abundant dissolved iron, was necessary for its formation. The nature of the deposits indicates that the layers of iron oxide were deposited by precipitation, possibly mediated by bacteria. And chemical analyses of the BIFs have revealed that the source of the iron was under-sea hot springs. As happens today in volcanically active regions of the oceans, seawater penetrated into the rocks of the seafloor along cracks and fractures and was heated to high temperatures, and it leached out iron as it circulated through the rocks. Because of seawater's low oxygen content, much of the dissolved iron remained in solution when the hot springs debouched on the seafloor. But if both seawater and the atmosphere were low in oxygen, how was the iron oxidized?

One possibility is that there were patches of early oxygen-producing photosynthetic bacteria in the upper sunlit zone of the oceans. When water from deeper levels, rich in dissolved iron, encountered this oxygen, the iron precipitated. This jibes with evidence that BIFs formed in relatively shallow water, and also with recent data from 2.6-billion-year-old South African rocks indicating that shallow ocean water was oxygenated while the deep ocean remained oxygen-poor. Although there is no definitive evidence for the presence of photosynthesizing cyanobacteria in the oceans during the Archean eon, there are bacteria living today that oxidize iron directly as part of their metabolic processes, without producing free oxygen, and it may be that similar organisms mediated the formation of BIFs.

Regardless of how the iron was oxidized, the conclusion from the BIFs that the early Earth's atmosphere contained little or no oxygen has been corroborated by other kinds of evidence. One of the most compelling clues comes from a chemical signature contained in small grains of pyrite (the mineral known as "fool's gold") in ancient sedimentary rocks. If that sounds a bit obscure, it is, but the tale is so instructive it is worth describing in some detail. It illustrates just how sophisticated the analytical tools for decoding the past have become.

In an oxygen-rich atmosphere like we have today, sunlight breaks down some of the atmosphere's oxygen (O_2) molecules, freeing up single oxygen atoms, which then combine with other O_2 molecules to make the three-atom molecule ozone (O_3). This is the process that forms an ozone layer in the stratosphere, and because ozone molecules absorb ultraviolet radiation from the Sun, the layer acts as a kind of sunscreen for the Earth. With no oxygen in the atmosphere, there would be no ozone layer and ultraviolet radiation would bathe the surface. Thus an Earth-surface feature that could quantify the amount of ultraviolet radiation penetrating through the atmosphere would be a good indicator of the atmosphere's oxygen content.

Ever ingenious, geochemists have come up with just such a measure: the isotopes of sulfur in pyrite from ocean sediments. How does that relate to ultraviolet radiation? Sulfur, in the form of sulfur dioxide gas emitted from volcanoes, is a minor component of the atmosphere. In geological terms, the volcanic sulfur dioxide doesn't linger long in the atmosphere; it is soon washed out and transported to the oceans, and eventually deposited in the sediments. But if it interacts with energetic ultraviolet radiation while still in the atmosphere, it undergoes chemical reactions that produce a unique isotopic fingerprint. That fingerprint is preserved even when the sulfur cycles into the ocean and is transformed into iron sulfide—pyrite—in the sediments, and it is an unequivocal signal that ultraviolet radiation penetrated deep into the atmosphere. When geochemists analyzed pyrite from sedimentary rocks spanning a wide range of ages, they found the distinctive sulfur isotope signature in virtually all samples older than 2.45 billion years, but not in younger rocks.

This discovery is about as convincing a piece of evidence as it is possible to find that the Earth had no ozone layer before 2.45 billion years ago, and therefore that the atmosphere lacked oxygen. Exactly why oxygen increased abruptly then, and by how much, is not known. But by most estimates the change was several orders of magnitude—from close to zero to somewhere near 1 percent. The transition has been dubbed "The Great Oxidation Event." Oxygen still had a long way to go before reaching today's levels, and the evidence shows that it didn't increase smoothly and steadily; there were bumps and dips along the way. But a threshold was crossed 2.45 billion years ago, roughly at the end of the Archean eon. From that time onward the atmosphere contained oxygen, even if the amounts were initially quite small.

The early Earth's atmosphere may have been different from today's in other ways too. In the early 1970s the astronomer and geoscientist Carl Sagan pointed out that 4.5 billion years ago the Sun's heat output would have been 20 to 25 percent lower than it is today, and that it has gradually increased to the present value. This conclusion, based on well-known details of how stars like our Sun evolve, has significant implications for the Earth's climate. It led to what has come to be known as "the faint early Sun paradox": if the Sun's energy output was so low in the distant past, why isn't there evidence in the geological record for a planet in deep freeze? The answer is almost certainly that heat-trapping greenhouse gases in the atmosphere were present at much higher concentrations than today, keeping the Earth warm.

The greenhouse gas we hear most about is carbon dioxide. Its concentration in the Earth's early atmosphere was probably several times today's level, but quite likely methane, an even more potent greenhouse gas, was also important for keeping the Earth from freezing. Like carbon dioxide, methane is a component of volcanic gases. However, most of the methane that makes its way into the atmosphere today originates from methane-producing microbes. It is not known when these organisms evolved, but they have an ancient lineage—they are members of the Archaea, the earliest known microbes. They thrive under oxygen-poor conditions, and once they had appeared on Earth they would have been a prolific source of the greenhouse gas.

On the Earth today, methane is destroyed via oxidation reactions; the average lifetime of a molecule in the atmosphere is less than ten years. But under the oxygen-free conditions of the early Earth, methane molecules would have remained in the atmosphere a thousand times longer, and overall concentrations could therefore have built up to much higher levels than exist now. In combination with the high concentration of carbon dioxide, that would have kept the planet from freezing and may even have made it quite warm. When free oxygen appeared in the atmosphere, the methane content would have begun to decrease, but even then the drop was probably not precipitous because oxygen remained low through most of the Proterozoic eon.

The oxygen-poor, greenhouse-gas-rich atmosphere of the Earth's first two billion years affected everything from ocean chemistry to biological evolution, climate, and the weathering of rocks on the Earth's surface. But a very different atmosphere and very different oceans weren't the only contrasts with today. The solid Earth, especially the nature and topography of the continents, was probably quite different too. The evidence for this is, if anything, even more elusive than that for the atmosphere. Many of the clues come from research into the beginnings of plate tectonics, the primary geological process affecting the continents today.

Surviving pieces of the Earth's oldest continental crust are small in size and number and are invariably highly metamorphosed, but they occur within all of today's continents (see figure 20 on page 129 for a map showing their distribution). Most of these small enclaves of ancient crust contain rocks that are, broadly speaking, granitelike. We have already seen that the Earth's oldest mineral grain, the 4.4-billion-year-old zircon crystal from Western Australia, solidified within a granitelike rock. Today the formation of granite is closely tied to plate tectonics, so this could be taken as indirect evidence that the process has operated since the Hadean eon. This is still a contentious issue, and opinions differ widely among geoscientists about just when modern plate tectonics began, but most now agree that the process was in operation by the end of the Archean eon, 2.5 billion years ago, and probably earlier. Recent research has uncovered indications that plate tectonics, or at least a close approximation of present-day plate tectonics, played a role in shaping the Earth's surface more than three billion years ago, and possibly as early as the Hadean.

How did early plate tectonics shape our planet's surface? Were there mountains and valleys like today, or were conditions quite different? The rock record, unfortunately, is not helpful for understanding ancient topography, partly because so little crust from that time still remains, and partly because the rocks that have survived simply don't contain any clear clues about the shape of the Earth's surface—or at least none that geoscientists have so far been able to decipher. However, some things can be inferred from theory and geophysical modeling. For example, we know that the Earth was much hotter during the Hadean and Archean eons than it is today, both because of its violent, heat-filled birth and because much more heat was produced by natural radioactive decay (radioactive heating has diminished as the Earth's natural radioactive isotopes have decayed away). Simulations indicate that under these hotter conditions, the lithosphere—the Earth's rigid outer layer—would have been much weaker than it is now.

With a weak lithosphere, great mountain chains like the Andes and the Himalayas could not have been supported. Our planet was likely much flatter than it is today; mountains would not have risen more than a few thousand feet, lower than the altitude of Denver. The subdued topography means that erosion, which is most intense in regions of high relief, may not have been as effective as it is today at breaking down the igneous rocks of the small existing continents and transforming them into sediments. However, the high level of carbon dioxide in the atmosphere, which would have produced acidic precipitation and therefore more intense chemical weathering of surface rocks, may have partly counterbalanced the topographic effects.

You have undoubtedly noticed the frequent use of "may have" and "perhaps" in this chapter. This reflects how difficult it is to be certain about conditions that existed and events that occurred billions of years ago. Closer to the present, the geological evidence becomes less ambiguous. Still, even though there is much we don't know about the Hadean and Archean eons, a great deal has been learned, as I hope this chapter has made clear. Evidence stored in rocks, coupled with inferences from theory, show that as the Archean eon drew to a close, the Earth's small continents were topographically subdued, there was almost no oxygen in the atmosphere, and the only life was single-celled organisms that inhabited the seas. No plants or animals graced the barren land, and average temperatures were high because carbon dioxide and methane in the atmosphere produced a strong greenhouse effect. Deadly ultraviolet radiation penetrated to the Earth's surface. Almost half our planet's history had passed by, and it was still a very different world from the one we know today.

CHAPTER FIVE

Wandering Plates

Before the 1960s geologists had some pretty bizarre ideas about how great mountain ranges like the Alps or the Andes formed. They were especially puzzled about the gigantic folds revealed by geological mapping in the Alps: thick layers of sedimentary rocks, originally deposited on the ocean floor but now thrust thousands of feet above sea level and looped over on themselves like a folded carpet. Unspecified "compressional forces" and vertical movements were evoked. Nobody could really say how these mysterious forces worked. Today, however, every geology student can tell you in great detail how plate tectonics is responsible for mountain ranges and many other aspects of the Earth's topography. The concept of a dynamic Earth, with thick plates moving about and interacting on its surface, has even provided everyday language with a vivid metaphor for radical change, with references to the "shifting tectonic plates" of the world order, or of politics, finance, or business.

Plate tectonics has been rearranging the face of the Earth over much of our planet's history. Its driving force is the cooling of the planet from its initially hot state, especially the cooling of the metallic core, a process that releases heat to the overlying mantle and fosters slow, large-scale convection. The steady plastic flow of the hot mantle rocks

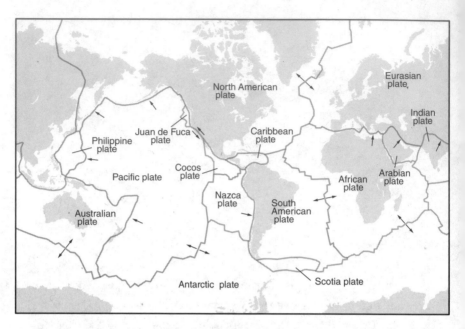

Figure 11. The major tectonic plates. Arrows show the relative motion between the plates. Most of the boundaries are either divergent, with plates moving apart at the ocean ridges, or convergent, with plates colliding at subduction zones. In a few places, such as western North America, two plates simply slide by each other along a fault. Many smaller plates are also recognized but are not shown on this map.

interacts with the cold, rigid outer skin of the Earth—the lithosphere— and causes it to move. The lithosphere is not a single, coherent layer, however, but is broken up into the many separate fragments that consti- tute the plates of plate tectonics (see figure 11).

The major tenets of plate tectonics were developed in the 1960s and revolutionized the field of geology. What had been largely a descrip- tive science evolved into a more predictive one, in which cause and effect were more easily connected. Plate tectonics became a model for understanding how the Earth works on a large scale. Many things that had been only poorly understood became clear: how mountain ranges are built, why earthquakes and volcanoes are localized in certain parts of the Earth, and how the ocean basins formed. The theory also forced geologists to think globally and to view even local geological features in a wider context. Although the oceans and atmosphere are not directly part of plate tectonics theory, earth scientists began to incorporate them—and their interactions with the solid earth—into geological thinking. As the interconnectedness of geological processes became apparent, geology departments in universities across the country and around the world rebranded themselves as departments of "earth system science" or "earth and environmental science" in recognition of this new thinking.

Some of the ideas incorporated into the theory of plate tectonics have a long pedigree. Many people have contemplated the world map and noticed that Africa and South America look as though they would fit together like puzzle pieces if the Atlantic Ocean were somehow removed. But that seems impossible; how could you possibly get rid of thousands of miles of ocean? Early in the twentieth century Alfred Wegener, a German scientist, suggested that the continents might simply have drifted through the regions now occupied by the ocean. He called this process "continental drift," and in 1915 he published a book outlining his ideas. Wegener's theory was ridiculed by physicists and didn't ever gain much traction, mainly because there were flaws in his notion of how continental drift might have operated. But the geological evidence he amassed was undeniable. He pointed out, among other things, that there are geological features in South America that end abruptly at the coastline and are matched by similar features in Africa. Without the intervening ocean, they would be continuous. He also showed that past glaciation had simultaneously affected parts of India, Africa, and South America, and that if the three continents had been squished together at the time, the glacial features could have been produced by a single ice sheet. Although most scientists did not embrace Wegener's ideas, the reality of his geological evidence—and of many similar observations gathered over the years—meant that various incarnations of "continental drift" came and went in the geological literature right up to the time of the plate tectonics revolution.

The observations that ultimately led to the theory of plate tectonics came from examination of the ocean floor, largely through studies funded by the U.S. Navy. In the years after World War II, as the range and depth over which submarines could operate increased dramatically, the Navy became very interested in the details of undersea topography. Seagoing geologists were happy to oblige, because the Navy's largesse gave them an opportunity to explore a largely unknown frontier. What they found was that the seafloor is far from the quiet, monotonous place many had envisioned. Although there had been hints from earlier work, the new studies revealed in unprecedented detail that the ocean floor is a place of great mountains, huge rifts, deep trenches, and active volcanoes. Especially intriguing was a line of mountains that bisects the Atlantic Ocean floor from north to south, following the outline of the continents on either side. It came to be known as the Mid-Atlantic Ridge, and it is peculiar because it is characterized by a central valley, with rocky walls rising up on either side. Furthermore, it was discovered that the Mid-Atlantic Ridge does not end in the South Atlantic. It joins up with a similar feature that runs east into the Indian Ocean, around Australia, and then into the Pacific, where it extends northward to California, making it a continuous, global feature.

Topography was not the only thing oceanographers charted, however. They also examined many other properties of the seafloor, including its magnetic characteristics. It was already well known that on land the magnetic field is influenced by the local rock types; rocks with high iron contents, for example, typically have a strong magnetic signal. This made magnetic surveys a valuable exploration tool in the search for metallic ore deposits. Geophysicists had also discovered that in thick sequences of lava flows, the *orientation* of the magnetic signal was reversed in some of the flows. Because igneous rocks "freeze in" the characteristics of the surrounding magnetic field when they cool and solidify, these researchers concluded that the Earth's field must have reversed periodically in the past, with north and south poles switching positions. Lava flows that erupted during these intervals had acquired the reversed signal.

Large-scale magnetic surveys on the continents typically show chaotic geographical magnetic patterns because very different rock types are often closely juxtaposed in the continental crust. However, the first magnetic surveys in the oceans, carried out in the 1950s off the northwestern coast of the United States, produced very different results. They showed that seafloor magnetism is highly regular, with long, linear swaths of similarly magnetized ocean-bottom rocks arranged in a zebra-striped pattern. The geophysicists who did the work were surprised—and puzzled. They didn't understand how this pattern was produced.

However, its origin became clear as additional seafloor magnetic data were gathered. Especially important was information from surveys conducted across the ocean ridge system in the Indian Ocean and North Atlantic. There the magnetic patterns, like those in the northeastern Pacific Ocean, were regular, with linear stripes of similarly magnetized rocks. But the striking feature of these surveys was that the stripes paralleled the ridge, and were very similar on either side of it—each side almost a mirror image of the other (see figure 12). It was a *eureka!* moment in the earth sciences. In 1963 two earth scientists at the University of Cambridge in the United Kingdom, Frederick Vine and Drummond Matthews, published a paper in which they proposed an explanation for the magnetic patterns: they suggested that the central valley of the ocean ridges is actually a rift in the Earth's crust where magma continually wells up to create new seafloor. As fresh, iron-rich seafloor basalt crystallizes and pushes off to either side of the ridge, it acquires a magnetic signature imposed on it by the Earth's field at the time. Changes in the magnetic field—especially the periodic reversals of polarity—are frozen into the seafloor rocks and account for the zebra-stripe pattern. The idea that the seafloor was spreading apart at the ocean ridges had already been suggested, but it was the magnetic data that really cemented acceptance of "seafloor spreading" and, eventually, led to the theory of plate tectonics.

Seafloor rifts apart

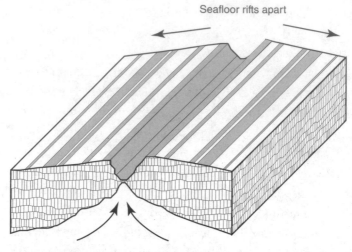

Upwelling magma from the mantle
forms new ocean crust

Figure 12. As new ocean crust forms at ocean ridges, it acquires a
magnetic signature from the ambient magnetic field at the time of
its formation; the result is a symmetrical pattern of magnetic stripes
on either side of the central rift. In this diagram, the dark stripes
represent seafloor with a magnetic polarity similar to today's, while
the white stripes signal seafloor created during times of reversed
polarity. The widths of the stripes vary according to the lengths of
the polarity episodes.

The entire seafloor is characterized by the same regular pattern of
magnetic stripes found near the ridges. Once this became clear, and
with Vine and Matthews's explanation of how the pattern is produced,
earth scientists realized that the ocean crust must be, geologically, relatively young. Observations like those made by Alfred Wegener decades
earlier also suddenly made sense. If new seafloor was forming along
the Mid-Atlantic Ridge and moving away, the Atlantic Ocean must be
getting wider. The youngest rocks are those along the ridge, the oldest at the margins of the ocean, close to the continents on either side.

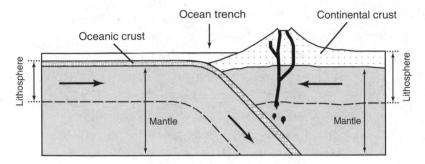

Figure 13. Cross-section of a plate boundary where oceanic and continental lithosphere collide, as, for example, along the western margin of South America. Arrows show the relative motion between the two plates. The oceanic lithosphere is being subducted below the continent, releasing water as it heats up. Melting begins in the hot mantle above the descending slab, and volcanoes erupt at the surface above the subduction zone. Material from the upper part of the ocean crust may be incorporated into the magma.

Run the clock backward and the Atlantic would close up, the seafloor disappearing back into the Earth's interior at the central ridge—and the common geological features in Africa and South America that Wegener had documented would be continuous. The continents hadn't drifted through the Atlantic basin as he thought; instead the ocean basin had formed when the continents split apart.

But there was an obvious problem. Unless the Earth is expanding, it would be impossible to make new, several-thousand-mile-wide ocean basins like the Atlantic. The only solution was that an equivalent amount of ocean floor must disappear somewhere else on the planet. In the plate tectonics cycle, that is exactly what happens: as new ocean floor is created along an ocean ridge, it is balanced by the destruction of an equal amount at a so-called subduction zone, where ocean crust plunges down into the Earth's interior (see figure 13). Frequently, but not always, subduction zones occur at the edge of a continent, and the seafloor descends beneath the continent as a continuous slab. Because there are large density differences between continental and oceanic crust—the continental rocks are much lighter—only ocean floor, not continental crust, is recycled back into the mantle in this way.

The tectonic plates, as noted earlier, are fragments of the rigid outer layer of the Earth, the lithosphere—a term that can be a bit confusing on first encounter, because this outer layer includes both the crust and the uppermost part of the mantle, as shown in figures 13 and 14. The base of the lithosphere is defined not by a change in chemical composition or rock type, but by a change in physical properties: the underlying material is close to its melting point and behaves more like soft plastic than solid rock, whereas the lithosphere is relatively rigid ("litho" means "rocky"). In contrast, the boundary between the crust and mantle is defined by a sharp change in rock type, a consequence of the melting processes that form the crust. The lithosphere averages fifty to seventy miles thick, but can be much thicker under continents and much thinner near the ocean ridges. The fragments of this layer that form the tectonic plates come in many shapes and sizes; they may contain continental or oceanic crust, or frequently both (figure 14). The boundaries between plates are marked by ocean ridges, where new lithosphere is being created; by subduction zones, where plates collide and one descends into the mantle beneath the other; and sometimes by faults, where two plates simply slide by each other. The relative motion between plates, whether they are colliding, moving apart, or slipping past each other, is slow, typically in the range of one to a few inches per year.

This sounds like a very simple and straightforward theory of what happens at the Earth's surface: a dozen or so major lithospheric plates jostling against one another, continuously getting renewed along ocean ridges and disappearing again down subduction zones. But why did the idea of plate tectonics revolutionize geology? How did it make it possible to understand processes that previously had no satisfactory explanation?

A complete answer to these questions would require more space than

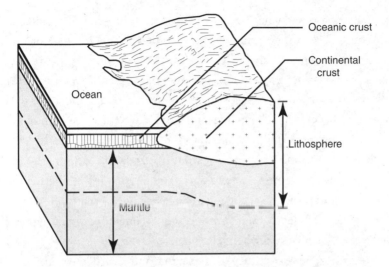

Figure 14. A simple sketch illustrating the relationship between mantle, crust, and lithosphere. Continental and oceanic crust are composed of different rock types and have very different thicknesses, and both can be part of a single lithospheric plate, as they are in this diagram (a real-life example is the eastern margin of North America). Compare the ocean-continent boundary in this diagram to the one in figure 13, where oceanic lithosphere subducts below a continent.

is available here, but it may be useful to explore a few examples. Take, for instance, earthquakes, which were obviously well known and studied long before the theory of plate tectonics was developed. Earthquakes occur when rocks are stressed to the breaking point, suddenly fracture, and move along a fault. Most earthquakes take place relatively close to the surface, where the rocks are cool, brittle, and susceptible to fracturing, but seismologists had long known that there are a few places—for example, along the western edge of the Pacific Ocean—where zones of earthquakes extend hundreds of miles into the Earth's interior. How this could happen was unclear, because at such depths mantle rocks were expected to be hot enough to flow under stress, not break with a great jolt. Plate tectonics provided the answer. The deep earthquakes occur at subduction zones, where cold, brittle ocean lithosphere descends into the mantle. Pressures on these great slabs of lithosphere are enormous, and although they travel slowly, they also heat up very slowly, remaining cool enough to fracture and generate earthquakes down to depths as great as four hundred miles or more. In fact, by mapping the exact locations of the deep earthquakes it is possible to trace the path of subducted lithosphere as it descends into the mantle.

A global map marking the location of earthquakes over a thirty-five-year period (figure 15) shows that they neatly outline the tectonic plates; although they are not entirely restricted to plate boundaries, earthquakes are clearly concentrated there (compare figures 11 and 15). Along the ocean ridges, where the plates are spreading away from one another, the zone of earthquakes is narrow and clearly defined; quakes occur at shallow depths in response to the cracking apart of the seafloor. In contrast, where tectonic plates are colliding, the earthquake zones are much broader because the tremendous compressional stresses resulting from the collisions are spread out over a wide area, and also because quakes occur throughout the subducted lithosphere as it angles down into the mantle. Within the wide band of earthquakes along the west coast of South America, for example, earthquake depths increase steadily from west to east, tracking the Pacific lithosphere as it plunges beneath the continent.

If you look closely at figure 15, you will notice that there are a few areas of high earthquake activity that are not at plate boundaries. In many cases, these can nevertheless be connected directly to plate tectonics; for example, many of the earthquake-producing faults in western and central China result from adjustments of the continental crust to a collision between India and the rest of Asia that took place many millions of years ago (more on this phenomenon later). In other areas, such as the eastern United States, a direct connection to plate tectonics is less clear. There earthquakes seem to occur along ancient zones of weakness in the crust that have somehow been reactivated, possibly partly in response to the general stresses transmitted through moving

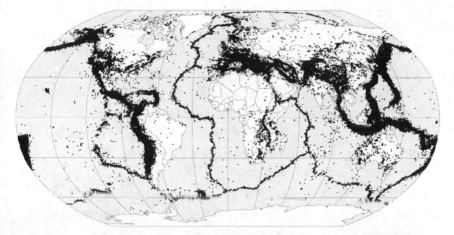

Figure 15. Locations of earthquakes that occurred between 1963 and 1998, shown as black dots. The vast majority took place along plate boundaries, as is evident by comparing this map with figure 11. Note that ocean ridges are marked by a narrow band of earthquakes, while the collisional plate boundaries are characterized by a wide band of seismicity. (Courtesy NASA; map created by Paul D. Lowman, Jr., and Brian C. Montgomery.)

and jostling plates. In yet other cases, such as the zone of earthquakes that runs down the eastern side of Africa, the seismicity signals a developing rift in the Earth's crust that may—many millions of years from now—split the continent apart and widen into a new ocean basin.

Most of the earthquakes distant from plate boundaries occur on the continents, but figure 15 also shows a dense, isolated cluster in the middle of the Pacific Ocean, far from the nearest ocean ridge or subduction zone. As you might guess, these earthquakes mark Hawaii; they are connected with the volcanic activity that occurs there. Rising magma beneath the volcano disrupts the surrounding brittle rocks, causing them to fracture and generating many small earthquakes. But this raises another interesting plate-tectonics-related question: Why is Hawaii there in the first place? Like earthquakes, most of the Earth's volcanism is restricted to the boundaries between tectonic plates. Yet the island of Hawaii, in the middle of the Pacific plate, is a massive volcanic structure, rising twice as high from its base on the seafloor to its peak as Mount Everest does, measured from *its* base to its peak.

One of the early proponents of seafloor spreading and plate tectonics, a Canadian geophysicist named Tuzo Wilson, was the first to understand the implications of Hawaiian volcanism. As an undergraduate student, I was lucky enough to take classes from Wilson when he was developing his ideas about plate tectonics. He was a big, charismatic man who spoke quickly and enthusiastically, and my initial encounter with him was in a first-year physics course. Homework problems for this course were onerous and there were many of them, but to lighten the burden we had weekly tutorials during which a graduate student, or occasionally a professor, would review the assignments with us. One day Wilson, who was a professor in the physics department, showed up at our tutorial. "Anybody can talk to you about the problems," he said. "I want to tell you about plate tectonics." For the next hour he held us spellbound; even though we got no help with the homework, it was by a long shot the best tutorial we had that year. Not long afterward I decided to major in geology.

Wilson realized that the volcanoes of the Hawaiian Island chain—a line of islands and submerged volcanoes that stretches some 1,500 miles from the "big island" of Hawaii toward the northwest and continues right up to the subduction zone bordering the Aleutian Islands—get progressively older toward the northwest. Only the island of Hawaii is currently active; all other volcanoes in the chain are dormant. Wilson proposed that each volcano was formed as the moving Pacific plate passed over a "hotspot" in the mantle, a fixed plume of hot material rising from deep in the Earth. The hotspot currently lies beneath the island of Hawaii, and as the Pacific plate moves gradually toward the northwest, today's active volcanoes will become dormant and a new island will form. That process is already under way. Oceanographers have discovered a large, active undersea volcano to the southeast of Hawaii. Its summit is still about three thousand feet below the sea surface, and it is not expected to emerge as a new island in the Hawaiian chain for tens of thousands of years. Formation of the volcanic chain has been compared to someone sending smoke signals from a campfire; each puff of smoke rises above the fire and is carried away on the breeze until there is a long, horizontal chain of discrete blobs of smoke in the sky. The physics of smoke signals and volcanoes are very different, but it is nevertheless an instructive image.

Hotspots are now a fixture of plate tectonics theory. The Hawaiian volcanic chain is perhaps the most dramatic example, but many others have been recognized. Especially in the Pacific Ocean, many islands are related to hotspots, including Samoa, Tahiti, and Pitcairn Island. In the Atlantic, Iceland is just one of several hot-spot-produced volcanoes. The phenomenon is not restricted to the oceans. Even Old Faithful, the famous geyser in Yellowstone National Park in Wyoming, can be traced back to a hotspot; as will be examined in more detail in chapter 11, the geyser sits within a gigantic volcanic crater, formed in a huge eruption 640,000 years ago, that was fed by magma from a hotspot.

As dramatic as hotspot-produced volcanoes can be, in volumetric terms the bulk of the Earth's volcanism occurs at plate boundaries, specifically along the spreading ocean ridges and above subduction zones. The character of volcanism in these two settings differs substantially, and—in an illustration of the power of plate tectonics theory as a framework for understanding how the Earth works—the differences account for the first-order feature of the Earth's topography: the vastly different average elevations of the ocean floor and the continental crust.

It comes as a surprise to many people to learn that most of the Earth's volcanic activity occurs deep in the ocean, unnoticed (except by earth scientists). As the seafloor cracks and spreads apart along the ridges that wind through the world's oceans, hot mantle rocks well up from below in response and begin to melt. The resulting lava is always basalt, a dense, dark-colored, iron-rich rock. The entire ocean floor, almost two-thirds of the Earth's surface, is made of this type of rock, with very little variation in its chemical composition from place to place. The ocean crust is quite thin, rarely more than six or seven miles thick, and it forms the top layer of the oceanic lithosphere. As the lithosphere spreads away from the ridge, it cools, becomes even denser, and settles deeper into the underlying plastic rocks of the mantle. As a result, the depth of the ocean increases steadily from a little less than a mile and a half at the ridge to around three miles in the deepest parts. And because of its high density, the ocean lithosphere eventually slides back into the Earth's interior at a subduction zone.

In contrast, rocks of the continental crust are much less dense. The processes that produce them are more complicated than those that form the ocean crust, but the starting point for making continental crust is again volcanism. In this case, however, the volcanism occurs at subduction zones. Why should there be volcanoes at subduction zones? As it turns out, when subducting ocean lithosphere descends into the mantle, it drags with it muddy seafloor sediments and ocean crust saturated with water. These components—water being the most important—mix with hot mantle rocks, lowering their melting temperature and initiating melting, just as salt spread on winter roads lowers the melting point of ice and promotes melting. Invariably, the subduction zone volcanoes form on the surface directly above the point where the subducting lithosphere reaches a depth of around one hundred miles, apparently a critical depth for the melting process. The lava that forms subduction zone volcanoes contains smaller amounts of heavy elements like iron than the ocean floor basalts, and larger quantities of light elements like silicon and aluminum. Thus over the long run of geological history, the operation of plate tectonics, reflected in the contrasting types of volcanism that occur at ocean ridges and at subduction zones, has had a profound effect on our planet: it has led to the formation of two very different types of crust, the low-density, long-lived crust of the continents, too light to be subducted, and the dense basaltic crust of the relatively ephemeral ocean basins, which is continuously recycled back into the mantle at subduction zones.

This brief description of how the Earth's crust has formed is somewhat simplified, particularly with regard to the continental crust. Not all continental rocks are made up simply of subduction zone lavas; many of them have been remelted and reprocessed during episodes of mountain building or as a result of other geological phenomena. The final product is a continental crust containing a wide range of different rock types. As pointed out in chapter 2, a good analogy is a large-scale refining operation that produces an end product very different from the starting material. Over geological time, these processes have resulted in a continental crust that is typically twenty to twenty-five miles thick, many times the thickness of the ocean crust. The low density of continental rocks allows them to "float" high above the dense ocean crust, which sinks deeper into the solid but yielding mantle.

Because continental crust is not subducted, it is, on average, much older than the ocean crust. The oldest parts of the seafloor occur along the margins of the Atlantic and Pacific Oceans; they are "only" about 200 million years old, whereas many parts of the continents are billions of years old. One of the implications of the crust-forming processes described here is that the continents must have grown in volume over time, which was certainly true in early parts of the Earth's history. However, even though it can't be directly subducted, continental crust can be eroded. Major mountain ranges have been thrust up and then worn down in the Earth's past. Eroded material mostly ends up as a layer of sediments on the seafloor; when ocean lithosphere reaches a subduction zone, much of that material is carried down into the mantle (some, however, is scraped off and thrust back up onto the continents, which is why intrepid climbers have found marine fossils at the very top of Mount Everest). Just as subduction zone volcanism adds to the volume of the continental crust, erosion and subduction reduce it.

In addition to helping earth scientists understand present-day geological processes such as volcanism and seismicity, the plate tectonics theory radically changed the way geoscientists view the past. Before plate tectonics was understood, very large-scale horizontal movements of parts of the Earth's crust, of the sort that Wegener envisioned in his theory of continental drift, seemed impossible. Now it is clear that they are a natural outcome of plate tectonics. Not only does this explain why rocks in Africa and South America, separated by thousands of miles of ocean, appear to be part of the same geological feature, but it also makes it possible to understand the juxtaposition of rocks of very different ages and character in the continental crust. Plate tectonics continually rearranges the geography of the Earth's oceans and continents, so that two pieces of continental crust now joined together might have been formed on opposite sides of the globe. Author John McPhee's popular book *Assembling California* describes how the complex geology of western North America is made up of many different small pieces of continental crust, formed initially at far-flung locations and brought together through the workings of plate tectonics.

How does this happen? One way to answer that question is to examine a classic example, the collision between Asia and India (obviously this involves two large landmasses, not small fragments as in western North America, but the principle is the same). If you could view the world of about ninety million years ago (a reconstruction is shown in figure 27 on page 191), you would find India as a small island continent far to the south of Asia, just off the coast of Africa. A wide ocean lay between India and Asia. But the ocean lithosphere was moving northward and being subducted under Asia, and India, on the same plate, was being carried along with it. Eventually, about fifty million years ago, the two continents began to collide. Too buoyant to subduct, the light rocks of the Indian continent simply crumpled against and interlayered with those of Asia, creating the world's greatest mountain range. Sedimentary rocks from the closed-up ocean basin were caught up in the collision, and some of them ended up high in the Himalayas—hence the marine fossils at the top of Everest. The two continents continue to push against each other to this day, and the stresses of the collision are responsible for a broad band of earthquakes across northern India and Tibet (see figure 15).

Reconstructing past plate geography is not always easy. The collision between India and the rest of Asia is well known because it occurred relatively recently. Plate movements can be tracked quite precisely back to the time of the oldest ocean crust, about 200 million years ago, using the seafloor magnetic stripe pattern as a guide. The pattern has been mapped throughout the world's oceans, and the ages of the stripes have been worked out from studies on land, where the periodic reversals of the magnetic field are recorded in sequences of well-dated lava flows. By matching the land records with the oceanic patterns, the exact age of nearly every part of the ocean crust has been established, allowing geoscientists to reconstruct the size and shape of the oceans—and the relative positions of the continents—at any time over the past 200 million years. Animations of how the tectonic plates have moved around the globe over this time period are popular displays in science museums.

Beyond the age of the oldest seafloor, however, working out plate motions becomes much more difficult, although not impossible. With an understanding of how plate tectonics works, and especially with knowledge of the characteristic rock types and geological features that are produced in different plate tectonics settings, it is possible to make reasonable reconstructions of continental positions back into the Precambrian. Long, linear mountain chains, for example, form at subduction zones (e.g., the Andes), and even more impressive ranges (e.g., the Himalayas) result when two continents collide over a subduction zone. These mountain ranges are characterized by specific rock types, a property that often makes it possible to decipher the origins of even ancient mountain ranges in terms of plate movements. Using such reasoning, it has been shown, for example, that the Ural Mountains in Russia were formed by a continent-to-continent collision between Siberia and eastern Europe that occurred about 300 million years ago, and that the Appalachian Mountains in the eastern United States were formed even earlier in a series of collisions between North America, parts of Africa, and several smaller fragments of continental crust.

But how far back in time can we push these kinds of analyses? When, exactly, did plate tectonics as we know it today begin to operate? Such questions are hotly debated among geoscientists, and in the summer of 2006 the Geological Society of America sponsored a conference to address them. The meeting was held in Wyoming, and it included discussions both in the conference hall and in the field among ancient rocks, the venue most geologists prefer for testing their ideas. By the end of the meeting there was still no universal agreement, but the consensus had clearly shifted toward an earlier rather than a later start for plate tectonics. That shift has continued in the intervening years.

Because subduction is such a key element of plate tectonics, unambiguous evidence for ancient subduction would be a clear indication that plate tectonics was in operation. Some geoscientists have argued on theoretical grounds that subduction as we know it now could not have occurred on the much hotter Earth of the Hadean and Archean eons. Convection in the mantle would have been more vigorous then than now, the tectonic plates would have moved about more rapidly, and "warm" ocean lithosphere may have been too buoyant to subduct easily. Proponents of this view point to the fact that certain metamorphic minerals known to be produced in subduction zones have never been found in ancient rocks.

When rocks from the ocean floor are subducted, they are subjected to immense pressures, but they remain relatively cool—cool enough to fracture in a brittle way and generate earthquakes down to great depths, as we saw earlier. This means that metamorphism at subduction zones is of a specific type, referred to by geologists (naturally enough) as "low-temperature, high-pressure" metamorphism. One of the diagnostic minerals of this kind of metamorphism is jadeite, more popularly known simply as jade. Next time you see a beautiful jade necklace, or a deep green jade figurine in a museum, you can thank plate tectonics.

In addition to jade, several other minerals are also diagnostic of low-temperature, high-pressure metamorphic conditions, and assemblages of these minerals are found in localities around the world where subduction zones have operated in the recent geological past. Strangely enough, virtually all such occurrences are in rocks with ages of less than one billion years. No one is quite sure why this is; however, few geoscientists think this date marks the start of plate tectonics. There are, for example, rocks of Archean age that on geochemical and geological grounds appear to have been part of volcanic island arcs, features that are linked to subduction. Today these arcs—the Aleutians and the Mariana Islands are examples—form where two oceanic plates collide and one plunges beneath the other. There is also some tantalizing evidence from ancient zircon crystals that the low-temperature, high-pressure conditions characteristic of subduction zones occurred more than four billion years ago.

Mark Harrison, a geochemist at the University of California in Los Angeles, and his colleagues were working on zircon crystals separated from the Western Australian quartzites that have yielded the world's oldest zircons when they noticed that some of the grains contain tiny inclusions of other minerals. Using a specially designed instrument, they were able to analyze these minuscule inclusions (typically only about four one-hundred-thousandths of an inch across) and deduce the temperature and pressure under which each crystal formed. The results were surprising. Some of the very old zircons, with ages of between 4 and 4.2 billion years, had formed at pressures corresponding to depths between about thirteen and twenty miles. This is not very deep as subduction zones go, but the *temperatures* these researchers deduced from the inclusion data were much lower than expected for that depth, and were consistent with conditions in a subduction zone. Furthermore, some of the mineral inclusions contain water, so it seems that even more than four billion years ago, during the Hadean eon, cool, water-rich surface material was dragged down into the Earth, to the zone where the zircons formed. The process may not have been subduction exactly as we know it today, but it was an analogous phenomenon, one that processed and remelted existing crust.

In 2009 additional evidence that plate tectonics operated very early in the Earth's history was discovered by a group of geophysicists from the Massachusetts Institute of Technology (MIT) and the Canadian

Geological Survey. Using remote sensing methods, they imaged what appears to be a piece of subducted ocean floor lying seventy miles beneath the current surface, at the base of what is known to geologists as the Slave craton, a chunk of Archean continental crust in northern Canada ("craton" is the geological term used for a surviving piece of old, stable continental crust). The Slave craton contains some of the oldest rocks in North America, and the researchers estimate that the subduction took place about 3.5 billion years ago.

It seems likely, then, that some form of plate tectonics has been operating on the Earth over most of its history. Today the framework of plate tectonics provides insight into the workings of most parts of our dynamic planet. In addition to giving us a better understanding of volcanoes, earthquakes, and the topography of the Earth, it helps to explain such disparate features as why mineral and petroleum deposits occur where they do, why marsupial mammals abound in Australia, and even, perhaps, why some of the Earth's ancient ice ages occurred when they did. Almost every conceivable geological phenomenon, and certainly most of those discussed in this book, is influenced in one way or another by plate tectonics.

CHAPTER SIX

Shaky Foundations

California is known for many things; one of them is earthquakes. It isn't mentioned in tourist brochures or real estate promotions, but the state is home to about three-quarters of the entire estimated earthquake risk in the United States. With the advent of plate tectonics theory, it became clear why: a portion of the boundary between the Pacific and North American tectonic plates runs through the state. This particular part of the boundary is neither a subduction zone nor a spreading center, but a fault—the San Andreas Fault—along which the plates slide past each other (see figure 16). Barring any abrupt changes in plate motions, a sliver of western North America extending from the southern tip of Baja California to San Francisco and beyond will ever so slowly move northward toward Alaska along the San Andreas. None of us will be around to witness the deserts of Baja California transformed into temperate forests as this happens, because it will take tens of millions of years. However, it *will* happen, just as surely as India split away from Australia more than 100 million years ago and sailed north to eventually crash into Asia and push up the Himalayas.

Over the past few million years and probably for much longer, motion along the San Andreas Fault has proceeded at a rate of just over an

Figure 16. *Left:* The great scar of the San Andreas Fault cutting through California, as seen from the air about halfway between San Francisco and Los Angeles. The view is toward the southeast, with the North American plate to the left and the Pacific plate to the right. (Courtesy U.S. Geological Survey; photo by R.E. Wallace.) *Right:* The trace of the fault through California. The arrows indicate the relative motion of the plates on either side.

inch per year. "Inches per year" is a useful way to think about plate movements, but it is worth remembering that it is just an average, not the reality. In the most famous of all California earthquakes, the great San Francisco earthquake of 1906, streams, fences, pipes, and roads were offset by ten feet or more across the fault in a matter of seconds. In one locality the movement was measured at nearly twenty-five feet. Hundreds of years of "average" motion occurred instantaneously. But the movement was localized. The whole Pacific plate didn't lurch at once, and visible signs of ground displacement only appeared along about three hundred miles of the San Andreas; in other places the crust on either side remained firmly locked together.

Scientific studies in the immediate aftermath of the San Francisco earthquake resulted in an entirely new concept of how earthquakes occur, one that has influenced research and prediction right up to the present. Henry Reid, a geology professor from Johns Hopkins University, was the lead scientist for the official investigation of the earthquake, and he proposed that behavior known as "elastic rebound" was responsible. It is an appealingly straightforward theory, not least because it makes prediction feasible. But many geoscientists now consider the elastic rebound theory to be inadequate for understanding interactions among the complex systems of faults that typically exist in seismically active areas.

The idea developed by Reid and his team was that forces in the crust—they did not know the origin of the forces in those pre-plate-tectonics days—gradually built up strain across the locked San Andreas Fault until it could no longer be accommodated and the rocks on either side snapped by one another, releasing the stress. Because displacement only occurred along a portion of the fault, Reid thought that the process must act on one segment of the fault at a time. He proposed that it might be possible to predict the next earthquake by putting a line of fixed markers across the fault as a monitor of its movement. The gradual accumulation of stress would be revealed by slow distortion of the straight line of markers into a kind of elongated "S" shape, and when the buildup became excessive an earthquake would be imminent. One logical outcome of this theory was that earthquakes on a particular segment of a fault like the San Andreas should occur at roughly regular intervals and be of about the same size.

The elastic rebound theory has, until recently, been the basis for most efforts at predicting the probability of earthquakes along the San Andreas. Instead of using Reid's idea of a line of markers across the fault, researchers now use GPS measurements to monitor crustal movements. But an accumulating body of evidence suggests that earthquakes along the fault may be neither very regular nor similar in size, and that external mechanisms, even a large, distant earthquake, may be as important as the amount of built-up stress in triggering them. One of the complexities affecting earthquake behavior and prediction, especially along a plate boundary like the San Andreas, is that faults typically exist as part of an interacting network. An earthquake along one fault may increase the stress on another. In figure 16, for example, I have portrayed the San Andreas as a single narrow line, but in reality it is a broad zone characterized by numerous individual faults. Everywhere along these faults and throughout the zone of movement you can get a sense of the tremendous forces involved as the plates grind past one another just by looking closely at the rocks: they are broken and crushed and granulated.

Earthquakes along these numerous faults are frequent, but fortunately most of them are small, recorded only by sensitive seismometers and not perceptible to the average Californian. The large ones, when they happen, can be terrifying. There is a story, perhaps apocryphal but conceivably true, about a geologist who moved to Los Angeles from a seismically quiet part of the country. The man said he was looking forward to being around during a good-sized earthquake because he had been teaching about them for so long that he was eager to experience one. Then early one morning in 1994 he was unceremoniously thrown out of bed by the Northridge earthquake, which shook Los Angeles violently. That was enough; he decided he would be quite happy never to experience a large earthquake again.

What throws people out of their beds, spills books from bookshelves, and knocks over bridges and houses is the rapid, repeated jolting caused by seismic waves traveling through the solid Earth and along its surface. The Earth quite literally quakes. We usually think of waves traveling through water, but they also propagate in solids; if you have ever hit a very solid object—a hard rock, say, or a piece of steel—with a sledgehammer, you will have experienced the unpleasant sensation of waves traveling back through the hammer and along the bones of your arms. If your blow is hard enough, the waves might even make your teeth chatter.

Seismic waves originate at the spot where rocks on either side of a fault slip by one another, the so-called focus of the earthquake, which can be a very small region, a long segment of the fault, or almost anything in between. A large amount of energy is released and spreads outward; generally speaking, the longer the segment of the fault that slips, the more energy released. The focus may be deep in the crust or even (as at subduction zones) within the mantle, but the earthquake location normally reported is its epicenter, the point on the surface directly above the focus. During an earthquake, movement along a fault can be horizontal, vertical, or, more frequently, some combination of the two. The Northridge earthquake was so damaging partly because there was a substantial component of very rapid vertical motion; the seismic waves caused the ground in Los Angeles to move violently up and down as well as from side to side.

Because large earthquakes have the potential to do enormous damage and take thousands or even hundreds of thousands of lives, geoscientists have long sought ways to predict them accurately. Henry Reid's elastic rebound theory was an early attempt. But it was based on information from a single earthquake, and to truly understand how earthquakes work—a prerequisite for prediction—requires comprehensive information from many earthquakes in diverse geological settings. Even in seismically active regions, large, damaging earthquakes are relatively rare, so the historical record provides only limited data. That leaves the geological record, the clues stored in rocks, as the most important source of information. But unlike most other geological phenomena, earthquakes have not left a record that extends very far into the past, and deciphering the few bits of evidence that do persist is particularly difficult. Faults, large and small, active and inactive, are prominent features of most geological maps. They occur almost everywhere in the Earth's crust, and they record displacement of one section of the crust relative to another at every scale, from a few inches to hundreds of miles. Earthquakes must have accompanied each of those movements, but it is often impossible to determine when they occurred or how big they were; the most recent large earthquake along a fault may have obliterated the evidence of earlier, smaller events. The difficulty of extracting information about earthquakes from the geological record has pushed geoscientists to devise some very creative approaches.

A particularly imaginative idea is the "precariously balanced rock" technique. In 1996 geophysicist Jim Brune proposed that it might be possible to put limits on the intensity of past earthquakes along a section of the San Andreas fault by examining what he referred to as "precariously balanced" rocks in the Mojave Desert and surrounding regions. The rocks that Brune singled out are startling when you come across them (figure 17); they look as though they could be toppled over with a slight push. However, appearances are deceiving. These rocks actually require a substantial jolt to dislodge them, and the dry climate, with its prevailing slow rates of erosion, insures that they remain delicately balanced over long time periods. In principle, an assessment of the shaking intensity required to knock these rocks off their pedestals, coupled with dating studies to determine how long they have remained in their current positions, should provide a measure of the time span since the last destructive earthquake. It should also allow scientists to put an upper limit on the size of any earthquakes that have occurred since then. There are critics of this approach, because the balanced rock idea is essentially based on negative evidence: we observe only the survivors and don't know how many similar rocks were toppled during the largest earthquakes. Nevertheless, Brune and his colleagues have made enough measurements to provide clues to the size of the largest earthquakes over a time span of many thousands of years, something that is difficult to determine from most other kinds of geological evidence.

Obviously, the balanced rock technique can only be used where such rocks exist, which means its applicability is geographically limited. More commonly, geological evidence of earthquake occurrence is preserved as a disturbance in a sedimentary deposit—a displaced soil layer, a disturbance in lake sediments, or a sudden change in sediment type in a coastal lagoon. Geoscientists frequently use carefully dug trenches close to or across active faults to reveal a vertical profile of

Figure 17. One of the precariously balanced rocks studied by Jim Brune and his colleagues. This example is from the Fort Sage Special Recreation Management Area, along the California-Nevada border about forty-five miles north of Reno, Nevada. (Photo by Jim Brune.)

disturbances in such deposits. In favorable cases, multiple past earthquakes can be documented this way, and if the ages of the disturbed layers can be measured using a technique such as carbon-14 dating, earthquake recurrence intervals can be calculated.

However, predictions based on the geological record are not precise. They do not tell us that a large earthquake will strike in October 2018 or April 2020. Rather, they are usually phrased in terms of probability: there is a 50-percent probability that a large earthquake will occur sometime during the next thirty years. Still, even this kind of information is valuable for local authorities, and for individuals too: earthquake forecasts of this sort might not stop you from buying a home in a desirable location that just happens to be an earthquake zone, but they would likely encourage you to make sure that your home—and your children's schools—are as earthquake-proof as possible. And they might also convince you to buy earthquake insurance.

Perhaps the best way to explore the characteristics of earthquakes is to focus on a few examples, examining what we know about the way in which the earthquakes occurred, what their consequences were, and whether there are clues in the geological record that might have alerted us to the imminence of these events—or that might indicate the probability of a similar earthquake happening in the future. An interesting instance of the latter concerns an earthquake-in-waiting in the Pacific Northwest of North America. There are, of course, many future earthquakes waiting to happen, but the possibility that a very large earthquake could strike the Pacific Northwest attracted a large amount of media attention some years ago when scientists discovered clues pointing to massive past earthquakes in the area, even though none have been experienced in historical times. This story has received considerable attention, and I wrote about it in detail in a previous book, *Nature's Clocks*, so I will describe it here only briefly.

Concern about the earthquake potential of the Pacific Northwest began in earnest in 1987 when Brian Atwater, a Seattle-based geologist with the U.S. Geological Survey, published evidence that at least six very large earthquakes had struck the region over the past seven thousand years. Atwater found that the earthquakes had abruptly dropped sections of the coastline in Washington State below sea level; the events were recorded by things such as dead coastal vegetation—including whole "drowned forests"—killed when the plants' roots were suddenly flooded by seawater. In places Atwater also found layers of coarse sand abruptly covering the fine mud of the suddenly submerged coastal marshes. The sand, it turned out, had been deposited by tsunami waves associated with the earthquakes.

Atwater's work made headlines because the region was not considered to be especially seismically active. Earthquakes do occur, but most of them are small. The written record going back to the earliest settlers contains no mention of large, destructive earthquakes. But there is a subduction zone just off the coast—named the Cascadia Subduction Zone by geologists—which had alerted scientists like Atwater to the potential for large earthquakes. Though subduction zones are the sites of many of the world's most devastating earthquakes, for some reason the Pacific Northwest had seemed to be an exception. Perhaps, the reasoning went, this was due to the unusual nature of this particular subduction zone. The lithosphere that plunges into the Earth's interior there is part of the Juan de Fuca plate (see figure 11), a small, young plate formed at an ocean spreading center located not far off the coast. The plate, still "warm" from its formation at the ocean ridge and therefore relatively buoyant, slips under North America at a shallow angle rather than diving steeply into the Earth's mantle. This might somehow prevent it from "locking up" and building up high levels of stress, and therefore might explain the occurrence of small earthquakes as the plates slide by each other but the absence of really large ones.

However, the geological record uncovered by Atwater showed that there is nothing particularly unusual about seismic activity along the Cascadia Subduction Zone. Great earthquakes, with magnitudes of 8 or larger, have occurred there in the not-too-distant past. The most recent has been traced to January 1700, just slightly too long ago to have been recorded by settlers. It was responsible for some of Washington State's coastal drowned forests, and it was almost certainly the source of a well-documented tsunami that struck Japan at that time. Atwater's research was a wake-up call for the region, showing that recent experience is no guide to long-term behavior. Great earthquakes in the Pacific Northwest are inevitable, and a very large earthquake today along the Cascadia Subduction Zone would cause extensive damage in cities like Seattle, Portland, and Vancouver. It might also set off tsunami waves that would wreak havoc all along the coast. Unfortunately—in spite of a large amount of research done since Atwater's initial work—the geological record still does not provide a firm basis for forecasting when the next big one will strike. Past great earthquakes have occurred at irregular intervals, and the record only goes back six or seven thousand years. About the only conclusion that can be drawn is that there is a very high probability of another great earthquake occurring in the region within the next thousand years. More immediate warnings are likely to come from earthquake precursors, like rapid changes in elevation that signal a buildup of stress within the crust and lithosphere.

The size of earthquakes not recorded in written documents, such as the one in 1700, can only be estimated from evidence in the geological record—from the extent of displacement along a fault, for example, or from clues to the geographical reach of the earthquake's effects. The Pacific Northwest earthquakes uncovered by Brian Atwater can confidently be termed "great earthquakes" based on such evidence, including the fact that they generated large tsunamis. In the modern era, however, the strength of an earthquake can be gauged much more accurately using seismometers.

One familiar measure of earthquake intensity is the Richter scale, devised by the seismologist Charles Richter in the 1930s. Richter's method was intended specifically for earthquakes in California and was based on the response of a particular type of seismograph. The scale is logarithmic, which is to say that one unit on the scale represents a factor-of-ten change in intensity; an earthquake registering 6.5 is roughly ten times more damaging than one of 5.5. However, it turns out that Richter's original method is not very good for accurately measuring the energy released in the largest earthquakes, and it has been revised in recent years. Nowadays the Richter scale is rarely even mentioned; instead, reports usually refer to an earthquake's "magnitude." For practical purposes, however, there is little difference between Richter's original scale and the new version. An earthquake with a magnitude of 2.3 is still a small earthquake, one with a magnitude of 5.6 is still a moderate earthquake, and one with a magnitude of 8.5 is a very large earthquake indeed and is usually referred to as a "megaquake" or "great earthquake." Richter's scale and its successor are open-ended, although earthquakes registering higher than 8 are (fortunately) rare. According to the U.S. Geological Survey, the largest earthquake ever recorded—which occurred in 1960 along the subduction zone off the coast of Chile—had a magnitude of 9.5.

One of the most damaging earthquakes in recent years occurred on May 12, 2008, in Sichuan province, China. It measured 7.9 on the magnitude scale. China has a long history of large earthquakes and has an active national program to monitor faults and assess earthquake hazards. The country is also a focus for international research in seismology. Even so, the Sichuan earthquake surprised the experts because it occurred along a fault that was thought to be in little danger of slipping—or at least not in danger of slipping in such a major way. It was a reminder of how difficult it is to predict earthquakes accurately. As the devastating earthquake that hit Haiti in January 2010 also showed only too clearly, they do not always strike where we expect them.

The Sichuan earthquake occurred in central China, far from the nearest plate boundary. Most other Chinese earthquakes are similarly distant from a plate edge. In spite of this, plate tectonics is the best way to understand most of the country's seismicity. With a little imagination, the origin of China's earthquakes can be traced back almost 200 million years to the huge ancient continent of Gondwana, which was made up of all the present-day southern continents assembled together in a single landmass. Gondwana broke apart in stages, and about 130 million years ago India split off from what is now Antarctica and began to move slowly north toward Asia. As described in the previous chapter, that eventually led to a collision between the two continents about fifty million years ago. However, the collision didn't bring the Indian plate's motion to a complete halt; GPS data show that it is still pushing north into Asia at a rate of approximately two inches per year. Great faults run along the arc of the Himalayan front, marking the boundary between the plates, and the stress of the collision is periodically relieved by large earthquakes along these faults.

The stress is dissipated in other ways too. As it rammed deeper and deeper into Asia, India not only crumpled and lifted up the rocks of that continent into high mountains; it also shuffled large swaths of Asia out of its way, pushing them eastward along a network of faults that today crisscross southeast Asia and China. The phenomenon, a kind of slow-motion eastward extrusion, has been likened to squeezing toothpaste from a tube.

The Sichuan earthquake was a minor adjustment of the crust to that eastward extrusion, although it was not so minor for those directly affected. In the area where the earthquake occurred, rocks of the Tibetan Plateau, moving eastward as a result of the collision, push up against the hard crust of the Sichuan Basin (see figure 18). A network of faults lies along the boundary between these two kinds of crust, and there is a dramatic change in elevation—the drop from the towering mountains of the plateau to the low-lying Sichuan Basin is one of the steepest anywhere on Earth. Clearly, there must be major geological forces at work to keep these blocks of the Earth's crust so out of balance.

The Sichuan earthquake occurred on the Beichuan Fault, one of several faults marking the boundary between the two crustal blocks. Although other earthquakes have occurred along the Beichuan Fault in the historical past, none has been as large as the 2008 event. The city of Beichuan, which sits right on the fault, has been in existence for about 1,500 years yet has never been destroyed by a large earthquake. Prior

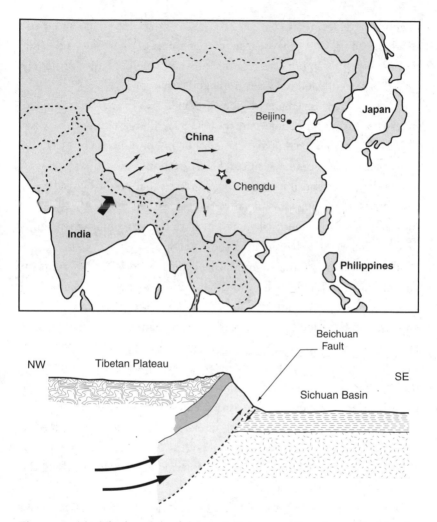

Figure 18. *Top:* The location of the 2008 Sichuan earthquake, marked by the star just to the northwest of the city of Chengdu. The small arrows indicate the movement of the crust as it is extruded eastward in response to the collision between India and Asia. *Bottom:* A cross-section showing the contact between the Tibetan Plateau and the Sichuan Basin along the Beichuan Fault. The small arrows indicate the relative motion across the fault; horizontal motion also occurred but is not illustrated here. (Based on an unpublished diagram by B.C. Burchfiel, L.H. Royden, and R.D. van der Hilst, 2008; used with permission.)

to 2008, there seemed to be little reason to worry too much about the Beichuan Fault; Chinese researchers spent most of their time and funds monitoring other, more active faults, most of which are also ultimately linked to the collision between India and Asia.

There were not even any obvious precursors to the 2008 earthquake. After the fact, some people said they had seen large numbers of toads shortly before the earthquake struck, but in spite of much anecdotal evidence, researchers have never uncovered any clear links between animal behavior and impending earthquakes that might be useful for prediction. It is quite possible that the toads in Sichuan were just an indicator of spring, and there may not have been any more toads hopping around in 2008 than in any other year.

The Sichuan earthquake emphasized that there are still significant gaps in our knowledge of how and why some earthquakes occur. One reason the Beichuan Fault was thought to be relatively benign is that it appeared to be made up of a series of short segments. On the surface, at least, there was no evidence that slippage had occurred along more than one segment at a time during past earthquakes, and short faults don't generate especially large earthquakes. But visible surface ruptures from the 2008 earthquake stretched over more than 150 miles and spanned several of the separate segments, which are now recognized to be imperfect surface expressions of a fault that is continuous deeper in the crust. Large earthquakes are therefore much more likely along the fault than previously thought.

The region around the Beichuan Fault is now the focus of renewed scrutiny by Chinese and international geologists. Although the 2008 earthquake relieved stresses that had been building up over hundreds and perhaps even thousands of years, these researchers have discovered that it may also have *increased* the stress on nearby faults, priming them for further earthquakes. The Sichuan earthquake may thus have increased the probability of another magnitude 7 or greater earthquake along one of the other regional faults. This possibility, and similar findings in other earthquake zones, is part of the reason why some geoscientists question reliance on the elastic rebound theory for earthquake prediction.

Many experts have also warned that in the frenzy of reconstruction after the Sichuan earthquake, government authorities may not be heeding a lesson they should have learned from past earthquakes around the world: much of the damage from large earthquakes is caused not by shaking alone—something that can be mitigated by enforcing strict construction standards—but by specific local geological conditions. In the case of the Sichuan earthquake, those conditions include the possibility of landslides, an inherent hazard of the precipitous local topography. Particularly if earthquakes hit during the rainy season, they have the potential to destabilize steep mountainsides and send debris surging down into highly populated valleys. Reconstruction needs to take into account not just how new structures are built, but where they are located.

Even though the severity of the 2008 Sichuan earthquake was a surprise, residents of central China—and other seismically active parts of the world—are no strangers to earthquakes. But in the early 1800s the last thing inhabitants of the American South expected was to be knocked off their feet—literally—by shaking ground. Beginning in late 1811, a series of massive earthquakes struck like bolts out of the blue there, unprecedented and without warning. Fortunately, they occurred in a sparsely populated area and human casualties were limited. But the seismic waves were widely felt, and the rearrangement of the natural landscape was spectacular.

The earthquakes began in the early hours of the morning of December 16, 1811, and they continued sporadically for more than a year. Based on eyewitness accounts and the extent of the known damage, it is probable that at least three separate earthquakes measuring between 7.5 and 8 on the magnitude scale occurred between December 1811 and February 1812. Each one of these was probably as large as the 1906 San Francisco earthquake. Many more moderate but still significant earthquakes shook the region during and after that three-month period.

Collectively, they are known as the "New Madrid earthquakes," after the Mississippi River town of New Madrid, Missouri, which was heavily damaged.

Without radio, TV, or the Internet, news traveled slowly early in the nineteenth century. Nevertheless, detailed reports about the earthquakes emerged fairly quickly, because even then the Mississippi was a busy river, and New Madrid was a bustling port frequented by riverboats, which carried the news up- and downriver. Furthermore, the earthquakes were felt over a wide portion of the American South. Newspapers from the winter of 1811–12 are full of letters from correspondents describing their own observations or those of friends. As always, there was a large audience waiting to learn about the natural disaster.

Some of the reports may be exaggerated, but by and large they seem to be earnest attempts to describe exactly what happened. They have the ring of another era and are fascinating to read. A letter from a resident of West River, Maryland, reported in the *Pennsylvania Gazette* (a Philadelphia paper), describes the sound accompanying one of the large earthquakes in January 1812 as being "like that produced by throwing a hot iron into snow, only very loud and terrific." In the *Louisiana Gazette*, we learn from another correspondent that the sound accompanying the first earthquake was initially like the distant noise of "a number of carriages passing over pavement," but then became a much louder "subterraneous thunder." The same writer supplies many additional details; he describes the temperature at the time of the earthquake, the foggy haze in the atmosphere, the unusual weather that year, and then goes on to outline in detail the characteristics of several small aftershocks. "In noticing extraordinary events," he says, "perhaps no attendant circumstances should be deemed unimportant: This is one of that character, and a faithful record of appearances in such cases as these, may form data for science." I'm sure that if he were alive today, this man would be an avid blogger.

Other reports speak of pendulum clocks stopped by the shaking, or of church bells set ringing, chimneys toppling, great cracks opening up in the ground, small islands in the Mississippi disappearing entirely, and people becoming nauseous in swaying buildings. As to the cause of the earthquakes, many opinions were offered. Some people thought they might be due to "subterranean fires," others called on the action of "electric fluid" within the Earth. A long-distance connection with volcanic events in the Andes was suggested. One of my favorites, described in another report that appeared in the *Louisiana Gazette*, makes reference to a great comet that had been visible in the fall of 1811: "The Comet has been passing to the westward since it passed its perihelion—perhaps it has touched the mountain of California, that has given a small shake to this side of the globe." There are also the inevitable claims that the earthquakes were some kind of divine retribution. One writer declared that perhaps the earthquakes had nearly destroyed the town of Natchez (the first capital of the Mississippi Territory and later of the state of Mississippi) because, he said, it was a place of "wickedness and the want of good faith."

Even in the midst of the earthquake chaos, not everyone lost their sense of humor. A man traveling down the Mississippi reported: "At New Madrid . . . the utmost consternation prevailed among the inhabitants; confusion, terror and uproar presided." Yet in the midst of that terror and confusion, a resident of New Madrid wrote the following in a letter to a friend:

> One gentleman, from whose learning I expected a more consistent account says that the convulsions are produced by this world and the moon coming into contact, and the frequent repetition of the shock is owing to their rebounding. The appearance of the moon yesterday evening has knocked his system as low as the quake has leveled my chimnies. Another person with a very serious face, told me, that when he was ousted from his bed, he was verily afraid, and thought the Day of Judgment had arrived, until he reflected that the Day of Judgment would not come in the night.

Two hundred years after the devastating earthquakes, the New Madrid area is still a center of seismic activity, although present-day earthquakes are small. Yet the southeastern United States is far from any plate boundary, and there is no process equivalent to the crustal extrusion in China that could connect the New Madrid earthquakes to a plate collision. How, then, can they be explained? And is there a significant probability that large earthquakes will strike the region again in the future? Both these questions have been the focus of intensive research in recent decades, and most geoscientists think the answer to the second question is an emphatic "yes." State governments in the region have taken note and have formed an organization to educate the public about the hazard and to promote earthquake safety measures.

The education efforts were given an unexpected but perhaps unwelcome boost when a retired biologist, Dr. Iben Browning, predicted that there was a 50-percent chance of a destructive earthquake occurring at New Madrid during the first few days of December 1990. Browning had calculated that gravitational forces from the Sun and Moon would be at a maximum then and could trigger a quake. His prediction—predictably enough—drew immediate and intense media attention, and despite overwhelming skepticism from geophysicists, there was a frenzy of activity in response. Schools closed, shops shut down, and TV crews descended on New Madrid. A few very concerned residents of the town fled. But nothing happened, and the disappointed—or perhaps relieved—TV crews packed up their equipment and left. At Tom's Grill in downtown New Madrid, the demand for earthquake burgers, with a fracture down the middle, dwindled.

The Holy Grail of accurate, short-term earthquake prediction is still a very long way off, and the Browning affair illustrates why most geoscientists are wary about issuing forecasts. Even predictions with little scientific credibility can be blown out of all proportion. But organizations charged with monitoring earthquake activity do issue warnings in terms of probability. On the basis of recent research, the U.S. Geological Survey predicts that the probability of a large earthquake in the New Madrid area, similar to those of 1811–12, is 7 to 10 percent over the next fifty years. In other words, there is roughly a one-in-ten chance that over that time the region will be struck with an earthquake that is potentially very damaging. Sliding down the intensity scale to magnitude 6, the probability climbs to between 25 and 40 percent. Such statements don't sound nearly as alarming as Browning's prediction, and they don't bring dozens of TV news crews, but residents of the southeastern United States have every reason to take them seriously. Even a magnitude 6 earthquake can inflict extensive local damage—and though most Americans think of California as the earthquake capital of the country, the inhabitants of several states along the Mississippi are at about the same risk as people in San Francisco.

Because of their size and remoteness from a plate boundary, the 1811–12 New Madrid earthquakes have become a case study for within-plate earthquakes. Years of research have illuminated the geological setting and provided a framework for how and why they occurred. Still, there are other places in the world with similar geology that are not seismically active, and why this is so is unknown. In 1996 the authors of a long and detailed scientific article about the New Madrid earthquakes quoted Winston Churchill, who said of Russia: "It is a riddle wrapped in a mystery inside an enigma." In spite of much progress in our understanding of the New Madrid events, the riddle has still not been completely unwrapped.

What *is* known about the 1811–12 earthquakes is that they are related to an ancient rift valley, a great scar in the Earth's crust that lies far below the surface. The origins of the rift date back many hundreds of millions of years, to a time when the forces of plate tectonics almost tore the continent apart but didn't quite succeed. The rifting did, however, produce a wide valley and was accompanied by volcanic activity. That valley has been named the Reelfoot Rift.

There are no obvious signs of the Reelfoot Rift at the surface today. Its presence beneath the New Madrid region is known from geophysical remote sensing data, which show that it is buried beneath several miles of sedimentary rocks. The rift is about two hundred miles long and fifty miles wide (see figure 19). Most earthquakes recorded in the region fall within its confines, typically at depths between two and eight

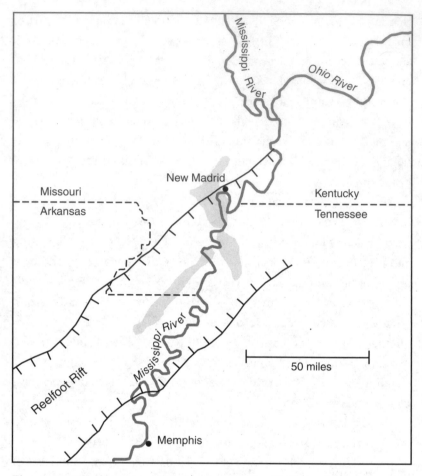

Figure 19. The location of the buried Reelfoot Rift, in the southeastern United States, based on geophysical remote sensing data. Most present-day earthquakes in the New Madrid area occur within the shaded regions.

miles beneath the surface, and the close correspondence between the earthquake locations and the rift have led most scientists to conclude that there is a cause-and-effect relationship. The town of New Madrid sits right on one edge of the buried rift.

But why should a feature in the middle of a tectonic plate and more than half a billion years old generate earthquakes today, or even two hundred years ago? One recent suggestion is that the rift was "reacti-vated" long after its formation when it drifted over a mantle hotspot about ninety million years ago. This intriguing idea came about when Roy Van Arsdale and Randel Cox, both at the University of Memphis, realized that plate reconstructions place the Bermuda hotspot, so named because it is believed to be responsible for the volcano that formed Bermuda, directly under the southern United States at that time.

Geologists have long known that this part of the North American continent was bowed up into mile-high mountains near ninety mil-lion years ago, and that there was accompanying igneous activity. Later, the mountains eroded away, the whole region sank below sea level, and several miles of ocean sediments accumulated on the submerged conti-nental crust. This uparching and later downwarping is consistent with passage of the region over a mantle hotspot.

Some geologists are skeptical of Van Arsdale and Cox's idea because they are doubtful about the reality of the Bermuda hotspot. There is no currently active volcano associated with the hotspot, and there is only a very indistinct track of extinct volcanoes tracing out the movement of the North American plate over it. However, the geological evidence for volcanism in the New Madrid area around ninety million years ago, as well as the uparching and subsequent downwarping of the crust, is undisputed, regardless of its cause. Such activity could certainly have reactivated faults associated with the original rifting. In more recent times, over the past few million years, the repeated loading and unload-ing of the North American crust by glaciers of the Pleistocene Ice Age may also have played a part in the reactivation. What is clear from both the large historical earthquakes and the present-day small ones is that faults in the Reelfoot Rift region constitute a weakness in the Earth's crust, and are prone to slip under sustained regional stress within the North American plate.

Recent GPS measurements show that stress in the Reelfoot Rift area is compressional—the rift is being squeezed from both sides. They also show that ground movements in the region have steadily decreased over the past twenty years or so, leading one group of researchers to suggest that the accumulated strain is decreasing and the earthquake risk diminishing. This is far from being the majority view, however. The U.S. Geological Survey stands by its 7 to 10 percent probability forecast for a large earthquake over the next fifty years. And the geological record shows that large earthquakes have struck the region repeatedly in the past after long periods of quiescence.

How do we know about these earlier earthquakes? The New Madrid seismic zone lies within the broad floodplain of the Mississippi River, and the entire region is underlain by thousands of years of accumulated mud, sand, and silt deposited during floods. When severely shaken— as happened during the 1811–12 earthquakes—this material undergoes liquefaction; it basically turns into a liquid slurry with little strength. Such behavior is common when earthquakes strike regions of loose, muddy, water-rich sediments; the collapse of cranes, docks, and other structures in the harbor of Port-au-Prince, Haiti, during the magnitude 7 earthquake of January 2010 was a result of liquefaction. During the New Madrid earthquakes, liquefaction produced "sand blows," mini volcano-like eruptions that shot slurries of sand and mud high into the air. One eyewitness claimed that he saw mud, sand, water, and coal thrown "thirty yards high." Deposits formed by the sand blows, sometimes several feet thick and spread over a diameter of up to a hundred feet, can still be found today, and when geologists began to map and date them, they found that some of these features greatly predated the 1811–12 earthquakes. Age determinations of the sand-blow deposits, virtually all of which occur within the area underlain by the Reelfoot Rift, show that at least four major earthquakes have struck the region over approximately the past four thousand years.

In spite of an immense amount of research into earthquake mechanisms and occurrences in places like the New Madrid seismic zone, along the San Andreas Fault, and in earthquake-prone regions such as Japan, China, and India, earthquake forecasting remains difficult. To date, only one specific prediction has been successful, in the sense that it was issued shortly before a major earthquake and allowed preparations to be made by local authorities. But while there were clear precursory signs that an earthquake might occur, the timing of the prediction may have been partly due to plain luck.

The prediction in question was made in China in 1975, and it concerned the area in and around the city of Haicheng, in the east of the country. Haicheng at that time had a population near one million, and the authorities were worried because there had been an increase in the number of small local earthquakes over a period of months. There had also been reports of changes in land elevation and water table levels. Then the frequency of small earthquakes abruptly rose, and the authorities decided to order an evacuation of the city. It was a brave move, because only a small percentage of large earthquakes are preceded by a rise in seismicity, but the very next day the city was almost completely demolished by a magnitude 7.3 earthquake. On such short notice, not everyone could escape, and more than a thousand people died. However, the toll would have been far higher had the warning not been issued.

Accurate prediction of the Haicheng earthquake may have been a lucky break, but it illustrates the importance of preparation for minimizing human suffering. Even if precise prediction of earthquake timing is not possible, simply knowing that there is a significant probability of a large event—knowledge that might be based on the geological record of past earthquakes, GPS measurements of ground motions, the number of small earthquakes that occur, or a number of other indicators—can be highly valuable. In places like Japan and California, it has prompted governments to impose strict building codes for earthquake safety and hold regular earthquake drills to familiarize the public and emergency services personnel with procedures to follow in the event of a natural disaster (most schoolchildren in California know exactly what to do if an earthquake strikes, even if their parents don't). When such measures are in place, loss of life, injuries, and physical damage to buildings and infrastructure are much reduced. If building codes are lax and preparation haphazard—a depressingly common state of affairs in many earthquake-prone parts of the world—even moderate to moderately large earthquakes cause widespread damage. In April 2009 a magnitude 6.3 earthquake near the Italian city of Aquila left nearly three hundred people dead and tens of thousands homeless. An Italian official commented afterward that a similarly sized earthquake in California would probably have caused minimal damage and no deaths.

Even if precise long-term earthquake warnings are not yet feasible, modern technology is being employed to give very short-term alerts that can help mitigate their effects. With networks of seismometers connected to high-speed computers, the initial seismic waves from an earthquake can be detected and analyzed, the epicenter determined, and an estimate made of the expected severity of shaking in nearby cities, all within seconds. Instantaneously, messages can be transmitted to subway systems, power plants, schools, hospitals, and emergency service personnel, advising them to follow prearranged procedures. Automated warnings can even be sent to every mobile telephone in the affected area. Earthquake waves spread out from the epicenter at speeds that depend on the local rock types, but are typically in the range of a mile to a few miles per second, so a warning might only precede arrival of the earthquake waves by seconds. Still, that could be enough time for schoolchildren to scramble under tables or subway trains to be halted. A short-term warning system is already in place in Japan; it sends messages to every school in the country when a large earthquake strikes. A similar system is being tested in California, where, depending on the location of the epicenter, the warning time could be considerable. For a large earthquake along the southern part of the San Andreas Fault, for example—something that has been forecast with moderately high probability—Los Angeles would have about a full minute's notice before the earthquake waves struck, time for many prearranged measures to be put in place.

Large earthquakes are a fact of life—or perhaps I should say, a fact of plate tectonics. When they happen, they remind us of the tremendous power inherent in the movement of lithospheric plates at the Earth's surface. Mother Nature can't be tamed, but earth science research enables us to anticipate her moves, and with this knowledge and careful planning it is possible to minimize the damage done during her periodic outbursts.

CHAPTER SEVEN

Mountains, Life, and the Big Chill

While reading the record of past earthquakes remains a tricky problem, the rock record is full of evidence about many other geological processes, even from very distant times in the past, as we have seen in earlier chapters. Here I will take up the journey through the Earth's history begun in chapter 4, and explore what evidence from rocks has revealed about the Proterozoic eon, the two-billion-year stretch of time between the Archean and the Phanerozoic eons. Even though the Proterozoic includes what some earth scientists have called the "boring billion"—a period of time when, in some respects, not much happened to the Earth's surface environment—the eon also witnessed momentous changes.

The aim of this chapter is not so much to give a comprehensive overview of everything that happened during the Proterozoic, but to focus on a few highlights. Rocks of Proterozoic age are much more common than those from the Archean, and for the most part the chemical and biological clues they contain are easier to decipher. For that reason, we can be more confident about reconstructing events from this part of the Earth's past. One of the most important aspects of research on the Proterozoic is that it provides insight into how the Earth behaved as a system under conditions very different from those of today. That not only gives us fundamental information about our planet's history, but it also enhances our ability to answer "what if" questions that can help to illuminate the future.

The name *Proterozoic* comes from the Greek *proteros*, meaning "first" or "former," and *zōē*, for "life." Early geologists placed the beginning of the Cambrian period—and thus the end of the Proterozoic—at the time when abundant fossils quite suddenly appeared in sedimentary rocks. Although Proterozoic rocks seemed to be barren, these scientists realized that there must have been primitive precursors to the fossilized organisms of the Cambrian; hence the name *Proterozoic* for the older rocks. There is no biological signal to mark the beginning of the eon, nor are there any layered sequences of sedimentary rocks where you can walk up and confidently place your hand on the boundary that separates the Proterozoic from the Archean, as can be done for most younger subdivisions of the geological timescale. The placement of the boundary is therefore somewhat arbitrary. In most cases, assignment of a particular rock formation to the Archean or the Proterozoic eon must be done on the basis of geological dating.

Most geoscientists, given half a minute or so to come up with what they consider to be the most important events of the Proterozoic, would list one or more of the following: the formation of large, stable continents; the evolution of eukaryotes and multicellular animals; the buildup of oxygen in the atmosphere and oceans; and severe glaciation during what are termed "Snowball Earth" events. There might be other things on their lists too, but these four are clearly the major events or processes that characterize the eon. All of them have been the subject of intensive research over the past few decades. A surprising and unexpected outcome of this work is the discovery that there may be close linkages among these seemingly very different phenomena.

How the first continents formed and grew on our planet is a topic that has been debated by geologists for more than half a century. Technological advances have made it possible to dig out clues that had not even been thought about just a few decades ago, and recent research has therefore had a huge impact on the debate. In particular, the ability to date ancient rocks with unprecedented accuracy, and to analyze minute mineral grains for chemical clues to past environments, has opened up whole new areas for investigation. As outlined in chapter 4, one of the fruits of that research has been the discovery of zircon crystals dating to the Hadean eon that—based on their chemical properties—must have formed in rocks not too dissimilar from those that characterize much younger continental crust.

The evidence from the zircons, coupled with the presence of 4.28-billion-year-old rocks on the shores of Hudson Bay, indicates that continental crust was present very early in the Earth's history. Small remnants of Archean-age "microcontinents" are incorporated into all of today's continents (see figure 20), but the total volume of these fragments is tiny compared to that of younger crust. An important question is whether this scarcity means that large volumes of early crust existed but were remelted and recycled so that we no longer recognize it, or whether there never was much ancient crust in the first place. The consensus view, based on many lines of evidence, is that early formation of continental crust was limited and the first continents very small. Until well into the Archean eon, and possibly into the early part of the Proterozoic, it is likely that no continents approached today's in size.

However, the geological record shows that the Earth's first supercontinents, truly gigantic landmasses that incorporated nearly all of the then existing continental crust into a single continent, appeared during the Proterozoic. Clues ferreted out of the rocks show that at least two of these appeared during the eon, and then fragmented into smaller pieces. More recently in the Earth's history, during the middle of the Phanerozoic eon, a third supercontinent assembled and then split up, and many geoscientists have concluded that long-term cycles of supercontinent formation and breakup are a natural consequence of plate tectonics.

The key to working out the existence and nature of the Proterozoic supercontinents, and how they formed and evolved, has been the merging

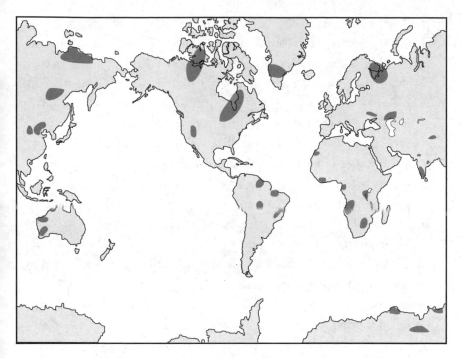

Figure 20. All of the present-day continents contain at least small fragments of ancient crust. The shaded areas on this map indicate the approximate locations of known Archean-age crust and remnants of early microcontinents. Small amounts of Hadean-age crust are present within a few of these regions.

of careful field studies with technology—in particular, measurements of the magnetic properties of rocks coupled with precise age determinations. The dating studies, crucial for understanding the chronology of supercontinent formation, have relied heavily on the silver bullet of geochronologists, uranium-lead dating of zircon crystals. Ubiquitous and highly resistant to having their internal clocks reset even by metamorphism, zircons are the mineral of choice for age determinations on very old rocks. Rock magnetism provides a completely different but no less important type of information.

In chapter 5, we saw that the magnetic properties of lava flows on the continents or the seafloor contain information about the polarity of the Earth's magnetic field when they formed. Other characteristics of the Earth's field are also captured by the magnetic minerals in various types of igneous rocks. One of these is orientation: at every point on the Earth's surface, the magnetic field has a slightly different orientation, from almost vertical to the surface near the poles to almost horizontal at the equator. Orientation is thus especially useful for determining the latitude at which a rock crystallized. With a few caveats, the magnetic properties of a continental rock can be used to determine quite accurately where on the Earth it formed—very useful information for reconstructing the past locations of continental fragments that have since traveled the world on their plate tectonic journeys.

Geoscientists who investigate rock magnetism are referred to as "paleomagnetists." The field of paleomagnetism got under way in the 1950s and early 1960s, before plate tectonics was understood, and at a time when most earth scientists assumed the continents had always occupied approximately their present positions. Then the early paleomagnetists made a startling discovery. From measurements of magnetic orientation in continental rocks of different ages, they deduced that the north magnetic pole must have moved about in the past, often to positions far from its present-day location near the Earth's rotation axis. This seemed odd, because theories of how the Earth's magnetic field is generated indicate that the magnetic poles and the rotational axis should always approximately coincide. The paleomagnetists dubbed the phenomenon they had apparently uncovered "polar wandering."

More difficult to understand was the fact that rocks with the same age but from different continents sometimes gave quite different locations for the magnetic poles. Only partly in jest, other scientists began to refer to the paleomagnetists as paleomagicians. But it eventually transpired that there wasn't any problem with their measurements; the problem was the underlying assumption that the continents had remained fixed in place. Once plate tectonics was understood, the magnetic measurements made sense; the continents, not the poles, had wandered. Now the interpretation can be turned on its head and used to work out the past locations of the continents by keeping the magnetic poles approximately fixed. The magnetic poles do wander a bit—scientists have been tracking the north magnetic pole for hundreds of years and have watched it move steadily across the arctic islands of northern Canada—but averaged over time, they coincide with the rotational pole.

As is true for most efforts to extract information from ancient rocks, thorough field studies that clarify the geological context and careful selection of individual samples for analysis often mean the difference between success and failure for paleomagnetic studies. If there are signs that a rock has been buried and strongly heated, for example, its magnetic signature may have been altered or even reset. And if a rock's orientation has been changed by folding or tilting, that too has to be taken into account when figuring out its original latitude. Consistent results from multiple samples of the same rock formation, collected over a wide geographical area, add confidence that the measurements are reliable. Many different research groups have done paleomagnetic work on rocks from the Proterozoic taking these kinds of issues into consideration. Together, their results provide convincing evidence that bears on the assemblage of Proterozoic supercontinents.

The existence of these supercontinents was initially proposed based on geological mapping that showed rock formations of very similar character and of about the same age scattered across various continents. The paleomagnetic studies clarified the original spatial relationships among these formations by showing that some of them were once contiguous. Rocks from locations as widely separated as central Australia and western North America, or Africa and Scandinavia, have been linked in this way, and precise age determinations have corroborated the connections. The process is a bit like putting together a jigsaw puzzle, except that the picture is indistinct and changes over time. Sometimes there are multiple possibilities that satisfy the existing data. But it has become clear that two supercontinents assembled and dispersed during the Proterozoic, each of them incorporating most or all of the continental crust that then existed. Geologists have named them "Columbia" and "Rodinia" (the latter from the Russian for "motherland"). There are fragmentary clues that an earlier supercontinent formed near the end of the Archean eon and lasted into the early part of the Proterozoic, but the evidence is not yet conclusive.

The Proterozoic supercontinents were dynamic features. They did not simply form and remain static; they changed continually, even if the changes proceeded at a slow, geological pace. New pieces of crust were added as other fragments separated. Columbia, the older of the two supercontinents, consolidated over several hundred million years, beginning about 2.1 billion years ago, and reached its maximum extent around 1.8 billion years ago. For most of the next 300 million years, it contained nearly all of the world's landmass, but then it split apart into multiple large continents. Several hundred million years later, the dispersed pieces of Columbia gradually reassembled as Rodinia. That happened between about 1.3 and 1 billion years ago, and Rodinia stayed largely intact for another 150 to 200 million years before splitting apart again (see figure 21 for a time line of Proterozoic events).

At the heart of Rodinia was what we now know as North America— or at least a good part of it; the continent has grown in size since it broke away from Rodinia. North America has long been viewed as a "classic" continent and has been central to the development of ideas about how continents are made. What is striking from a geological standpoint is that (in a slightly simplified view) the rocks that make up North America get progressively younger toward its margins. The oldest rocks, those of the so-called Canadian Shield, lie at its center and are surrounded by belts of younger rocks. Somehow, the continent must have grown from an initial small core by the addition of new crust around the edges.

With the discovery of plate tectonics, it was realized that this growth took place at subduction zones. Two related processes were responsible, both involving volcanism: the first added new volcanic material directly to an existing continent, as volcanoes in the Andes do today; the second built offshore volcanic island arcs that were later plastered on to the edge

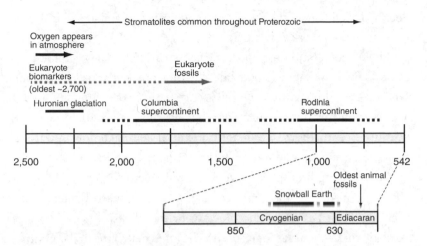

Figure 21. A time line of some of the important events and features of the Proterozoic eon. Dates are given in millions of years before the present. The Cryogenian is shown here for ease of comparison with events described in the text, but it is not a formal subdivision of the geological timescale.

of the continent through plate collisions. The Appalachian Mountains, stretching from Newfoundland in Canada almost to the Gulf of Mexico, were formed by both of these processes, beginning nearly 500 million years ago, when a subduction zone lay along the entire eastern margin of North America. This was long after the breakup of Rodinia, and it was one of the steps in the extended formation of the later Phanerozoic supercontinent, Pangea.

Just inland from the Appalachians is a parallel but much older belt of rocks that was formed in a similar way during the assembly of Rodinia. Known to geologists as the "Grenville belt," it too marks an ancient subduction zone along which continents and island arcs collided with the core of North America and remained attached. The Grenville rocks were formed between about 1.1 and 1 billion years ago, when the colliding continents thrust up towering mountains that resembled the Himalayas or Alps, although you would not know it today. All that remains is an eroded, flat-lying landscape of metamorphic rocks that were once the deeply buried core of the mountain range. Erosion and geological time have done their work very efficiently.

The agglomeration of supercontinents through plate tectonic processes is fairly well understood, but why they break apart is less obvious. One theory, although not the only one, is that very large continents act as a kind of thermal blanket, causing the mantle underneath to heat up. Eventually the hot mantle material begins to flow upward as a plume or hotspot due to its buoyancy, causing uplift and stretching of the overlying lithosphere. Cracking and fracturing ensue, together with volcanism, and the continent begins to rift apart. Evidence that such a process has occurred comes from the widespread presence of volcanic rocks known as "dikes" with ages that match the breakup times of the supercontinents. The dikes are narrow, vertical sheets of lava that cut through the continental crust along fractures. These fractures were conduits for the large volumes of lava—now mostly eroded away—that flooded onto the rifting continent. If Africa splits apart along the East African Rift, swarms of dikes will be evident along the rifted continental edges tens or hundreds of millions of years from now.

Regardless of exactly how the Proterozoic supercontinents broke up, both they and the smaller landmasses that formed from their breakup were very different in appearance from today's continents. About the only similarity was their topography, with high mountain ranges, flat plains, and deserts. But they hosted no life. Steep topography, the lack of vegetation, and a high content of carbon dioxide in the atmosphere, which produced strongly acidic precipitation, combined to erode the surface rocks intensely. That conclusion is based firmly on evidence from Proterozoic sedimentary rocks, which are characterized by extensive, thick beds of pure quartz sandstone, a signature of continental erosion.

In contrast to the barren continents, though, the oceans were endowed with abundant life during the Proterozoic. Bacteria and archaea, the simple single-celled organisms that may have arisen as early as the Hadean, were the dominant life forms. Stromatolite structures formed by these organisms are common in Proterozoic rocks; they are the poster fossils of the eon. The first true fossils of eukaryotes, with cell nuclei and other complex features, appear at about 1.8 billion years ago. (As described in chapter 4, eukaryote-specific biomarkers have been found in older, Archean rocks, suggesting that they originated much earlier.) Toward the very end of the Proterozoic eon, a whole ecosystem of more complex, multicellular animals appeared—and almost simultaneously the stromatolites declined in abundance. Intriguingly, the evolution of life during the Proterozoic, the supercontinent cycles, and changes in climate and the chemical makeup of the atmosphere may all be linked.

A connection between life and the atmosphere, especially its oxygen content, is readily understood. The rise of oxygen during the Protero- zoic must be tied to the evolution of green plants, which are the source of atmospheric oxygen through photosynthesis. But what evidence is there for connections among supercontinent formation and breakup, Snowball Earth episodes, atmospheric composition, and the evolution of life?

Perhaps the best way to answer this question is to examine chrono- logically how different parts of the Earth system interacted during the Proterozoic. Simultaneity, or near simultaneity, doesn't always prove relatedness, but when the geological record shows that major changes in different parts of the system occurred at more or less the same time, it is reasonable at least to investigate the possibility that the events are linked. And if the chemistry and physics behind the changes are com- patible with a linkage, then there is a strong possibility that there is indeed a close connection.

I should emphasize that there is still some uncertainty about exactly how the Earth's history unfolded during the Proterozoic, especially dur- ing early parts of the eon. There are gaps in the rock record, and many of the surviving rocks have been metamorphosed. The version I give here is subject to revision, but it is plausible given the evidence we have today. To set the stage, it is necessary to begin by looking again at condi- tions late in the Archean eon, shortly before the Proterozoic began.

Some Archean sedimentary rocks have chemical properties suggesting that the oceans were quite warm, possibly over fifty degrees Celsius (122 degrees Fahrenheit). If the sea was that hot, land temperatures would have been even higher. This is the crux of the faint early Sun paradox mentioned in chapter 4: during the Archean, solar energy incident on the Earth was only 75 to 80 percent of its present value, and by that measure the surface temperature should have been lower, not higher, than today. The only plausible solution to this problem is that carbon dioxide and methane concentrations in the atmosphere were high, keeping the Earth comfortably warm—or maybe even uncomfortably hot—through the greenhouse effect. As we saw earlier, methane probably built up to very high levels under the oxygen-free conditions of the Archean atmosphere. But there was a delicate balance at play—if methane levels rise too high, a methane haze starts to form. The idea of an atmospheric fog of methane is not science fiction; Titan, the largest moon of Saturn, is shrouded in just such a haze. On the Archean Earth, if methane levels got high enough, the pinkish-brown methane fog would have countered the greenhouse effect by blocking sunlight.

In South Africa, geologists have discovered a type of sedimentary deposit known as "tillite," a hodgepodge of small and large rocks and fine sediments characteristically formed by glaciers. The South African tillite is 2.9 billion years old and is the only such deposit known from the Archean eon. Boulders from the tillite show the scratches and scouring marks of glaciers, and the deposit has thus been interpreted as evidence for the Earth's first ice age. It has been suggested that high methane levels in the Archean atmosphere initially helped keep the Earth very warm through the greenhouse effect, but later brought on the glacial episode when a methane haze formed, cooling the Earth for a short period.

As far as we can tell, temperatures had risen again by the end of the Archean and into the early part of the Proterozoic eon. However, the Proterozoic was characterized by more than one glacial period, and several of these were exceptionally severe (see the time line in figure 21).

The first—which was actually a succession of closely spaced cold intervals—occurred between about 2.45 and 2.2 billion years ago, shortly after the eon began. It is known as the "Huronian glaciation," and it too may have been linked to methane in the atmosphere, although in a different way from its Archean predecessor.

The beginning of Huronian glaciation coincided closely with the Great Oxidation Event described in chapter 4, the abrupt rise in atmospheric oxygen around 2.45 billion years ago. Exactly why oxygen appeared in the atmosphere at this particular time is not known in detail, except that it must have been related in some way to photosynthesis and the evolution of green plants. The increase, although very large in relative terms because its starting point was essentially zero, only brought the oxygen content to about 1 percent or less of today's levels, but it was enough to sharply reduce the average residence time of methane molecules in the atmosphere (methane is destroyed by oxidation). As methane concentrations began to drop, so did the efficiency of the greenhouse effect, and temperatures declined.

The Earth was not continuously in an ice age throughout the entire period now identified with the Huronian glaciation; rocks from the Great Lakes area of Canada contain evidence for three well-defined glacial periods separated by warmer intervals, and it is likely that there were more. Rocks of about the same age from the United States, South Africa, and Finland similarly provide evidence for glaciation, indicating that the ice age was global in extent. The magnetic properties of some of the rocks associated with glaciation place them at relatively low latitudes, a sign that the cold was intense and possibly extended into the tropics. One or more of the Huronian glaciations may have been equivalent to the Snowball Earth periods that occurred later in the Proterozoic, near the end of the eon.

I have already mentioned several times that the geological record is not continuous, but it is worth emphasizing again. The layer-cake strata exposed in the walls of the Grand Canyon, for example, span a total of more than a billion and a half years of the Earth's history, but there are hundreds of millions of "missing" years, with no rocks to represent them at all. Just because the oldest rocks showing evidence for Huronian glaciation are 2.45 billion years old and the youngest 2.22 billion does not mean that there was no glaciation slightly before or after these dates. The exact timing of the Great Oxidation Event is similarly fuzzy. Nearly all rocks older than 2.45 billion years show that the atmosphere was devoid of oxygen, while rocks younger than about 2.32 billion years indicate that the atmosphere was oxygenated—but again, gaps in the record make exact dating of the transition impossible. Even so, as geological coincidences go, the overlap in timing between the Great Oxidation and Huronian glaciation is striking: oxygen rose and the Earth froze. Or perhaps it was the other way around; the dating is not precise enough to tell. But a reasonable interpretation is that the oxygen came first, reducing methane concentrations, which caused a decrease in the greenhouse effect and pushed the Earth into an ice age.

There is also circumstantial evidence that links greenhouse gases, the Great Oxidation, the Huronian ice ages, and the formation of continents. One of the reasons the Archean-Proterozoic boundary was placed at 2.5 billion years ago is that geologists found subtle changes in the nature of crustal rocks formed before and after this time. The changes didn't occur instantaneously, even in a geological sense, but the crucial observation is that the mineral makeup of Archean sedimentary rocks indicates that many of them were produced by weathering of basalt, while sedimentary rocks from the Proterozoic are more likely to be the product of weathering of granitelike rocks. This doesn't mean that granitic crust didn't exist before 2.5 billion years ago; we have already encountered evidence that subduction, or similar processes, formed granitic rocks much earlier. But the sedimentary rock data indicate that continental crust similar to today's only became the dominant source of sediments in the Proterozoic. This may reflect a gradual but fundamental shift in the mode of crust-building, from basalt-dominated volcanism, like that of Iceland or Hawaii, toward processes that resulted in the stabilization of granitic-composition continents through subduction-zone volcanism. That has implications for the greenhouse gas content of the atmosphere during those early times.

Why should that be? Basalt is the characteristic magma formed by direct melting of the mantle, and it brings with it dissolved methane and carbon dioxide from the Earth's interior, which is mostly released into the atmosphere. Processes that produce granitic-composition continental crust, on the other hand, are more complex, as explained earlier; overall, they transport much less of these gases to the surface. A decrease in the amount of basaltic volcanism, even if accompanied by increasing subduction zone volcanism, would therefore lead to a decrease in the greenhouse gas content of the atmosphere. It might also partly account for an increase in oxygen content, because less of the oxygen produced by photosynthesis would be used up in reactions with methane. The net result would be lower temperatures, possibly low enough to trigger the Huronian ice age.

Continent building, glaciation, and biological evolution may also be related in other ways. As continents increased in size and number and began to support more extreme topography, erosion and chemical weathering of surface rocks increased. Over the long term, chemical weathering is the main process that removes carbon dioxide from the atmosphere: carbon dioxide dissolves in rainwater to form carbonic acid, which attacks and dissolves rocks; it is then carried to the ocean by streams and rivers, along with other dissolved materials from the rocks, and eventually gets stored, more or less permanently, on the seafloor as a component of sedimentary rocks (see figure 25 on page 172). Thus continent building, especially mountain building at subduction zones, may have decreased the amount of carbon dioxide in the atmosphere and initiated cooling. Furthermore, the dissolved materials from increased weathering of the developing continents included an abundance of important biological nutrients like phosphorus and iron. Carried to the ocean, these nutrients may have stimulated the proliferation of life and played a role in the evolution of photosynthesizing cyanobacteria and consequent oxygenation of the atmosphere.

This is a somewhat speculative scenario; we don't know for sure that this is the way things really happened. But we do have a chronology of events, and we do know in some detail how the processes described above—destruction of methane by oxidation, extraction of carbon dioxide from the atmosphere through weathering, global temperature changes caused by changes in atmospheric greenhouse gases, and fertilization of the ocean with nutrients from increased weathering—operate and what the consequences are. At the moment, there is no evidence that contradicts the interconnections among these processes. But it is likely that the details will change and the picture will become clearer as geoscientists uncover new clues in the Proterozoic rocks.

The first few hundred million years of the Proterozoic were quite eventful, but from the end of the Huronian glaciation until nearly 800 million years ago the surface environment was relatively stable. This is the period that some geoscientists refer to as the boring billion (it is actually more like a billion and a half, but that doesn't roll off the tongue quite so nicely). These scientists tend to be geochemists who focus on the oceans and atmosphere: oxygen in the atmosphere remained low and fairly stable during this interval, and not much changed in terms of ocean chemistry. But the period also includes the assembly of Columbia, the first clearly defined supercontinent, beginning about 2.1 billion years ago, and the formation of the earliest large-scale linear mountain chains, similar to the Andes. It also saw the preservation of the first eukaryote fossils around 1.8 billion years ago.

It is difficult to overstate the importance of the evolutionary step that produced the eukaryotes. Biologists have called it one of the most important transitions in the evolution of life. Had it not occurred, you wouldn't be here to read this book. DNA evidence shows that eukaryotes carry genes from both types of organisms that existed before them, the bacteria and the archaea. Somehow, one variety of these prokaryote ancestors ingested or was invaded by the other, and instead of destroying each other, the two lived on in a kind of symbiotic relationship. This, apparently, is how the internal structures of eukaryotes, like mitochondria (tiny membrane-enclosed, energy-producing units) originated. The mitochondria—there can be thousands in a single cell—allowed the eukaryotes to grow much bigger than their prokaryote ancestors, because energy-generating reactions could be carried out internally by multiple mitochondria instead of through the external cell wall (as a cell gets larger, its volume increases much more rapidly than its surface area, and a prokaryote cell, without mitochondria, would quickly reach the point at which it could no longer generate enough energy to keep itself running). The presence of various internal structures like nuclei and mitochondria also allowed eukaryotes to form multicellular organisms and develop specialized functions in different cells. That set the stage for the evolution of complex animals and plants, though this would not happen for more than another billion years.

At about the time the first eukaryote fossils were preserved, approximately 1.8 billion years ago, Columbia reached its maximum size. As outlined above, the supercontinent broke apart a few hundred million years later, and the resulting smaller continents drifted around the globe until they began to come together to form the second great Proterozoic supercontinent, Rodinia, beginning about 1.3 billion years ago. Then, near the end of the eon, there was a great burst of geological activity. In (geologically) rapid succession, over a period of about 300 million years, Rodinia began to break apart and the Earth was gripped in a series of glaciations, two of which were intensely cold Snowball Earth episodes. At the same time, oxygen levels in the atmosphere, which had fluctuated during the boring billion but remained low, shot up almost to present-day levels, and there was a flourishing of complex, newly evolved animals on the seafloor. Once again we can ask whether there are connections among these seemingly unrelated events. The answer from most geologists who have examined the evidence carefully is a cautious "probably."

The final breakup of Rodinia into smaller continents happened between about 850 and 800 million years ago, and almost simultaneously the Earth's climate cooled drastically. The geological evidence for glaciation is so widespread that the interval between 850 and 635 million years ago is now called the Cryogenian. (The Cryogenian is not an official period of the geological timescale, but the term is in widespread use among earth scientists.) At least two ice ages during this time are thought to have been global in extent, with glaciers extending into the tropics, and it is possible that even large parts of the ocean froze over. The term *Snowball Earth* was conceived to describe these global ice ages.

Reading the rock record from 800 million years ago is much easier than extracting information from heavily metamorphosed Archean or early Proterozoic rocks, and as a result we know quite a bit about Snowball Earth. There is circumstantial but plausible evidence that as Rodinia broke up, weathering of the new continents withdrew enough carbon dioxide from the atmosphere to weaken the greenhouse effect and cause the Earth's climate to descend into the cold of the Cryogenian. Uniquely, all of these continents were at low or mid-latitudes (a similar arrangement has not been repeated since) and surrounded by relatively warm oceans. With a prolific source of moisture and mild to tropical temperatures, weathering—and the associated carbon dioxide withdrawal—was rapid. The unusual clustering of continents at low latitudes also had another effect. Land reflects more of the Sun's energy than water, especially if the land is barren of vegetation, as it was 800 million years ago. With all of the continents located within the zone where most of the Sun's energy falls, more of that energy was reflected back into space. And although the Sun's energy output had increased significantly since the earliest part of the Earth's history, it was still about 6 percent lower than it is today. All of these things conspired to lower temperatures globally and push the Earth into a prolonged period characterized by multiple ice ages, one or more of which were probably the most severe the planet has ever experienced.

The magnetic properties of rocks from the post-Rodinia continents were the crucial clues to their low-latitude locations during the Cryogenian. When it became clear that glaciation was widespread on all of these continents, some geoscientists were incredulous. What made the evidence even harder to swallow were signs that at least some of the glacial deposits formed near sea level, not on high mountains. There are low-latitude glaciers today, but they only exist at high altitudes in mountain ranges such as the Andes and Himalayas. During the Cryogenian, however, ice sheets scraped debris from the continents, flowed to the shoreline and beyond, and dumped jumbled glacial deposits directly into the shallow water of the continental shelf. How could that happen in the tropics? Once again there was suspicion that somehow the paleomagnetists had gotten it wrong and the continents had really been at high latitudes during the glacial periods. But in spite of hundreds of investigations seeking to find errors in the paleomagnetic measurements, no flaws have been discovered.

Some computer models show that glaciers on mid-latitude continents, together with sea ice around the poles, could reflect so much of the Sun's energy back into space that the Earth would be tipped into a runaway freezing mode. These models, of course, depend crucially on the assumed initial conditions, such as the concentration of greenhouse gases in the atmosphere in the late Proterozoic. We don't know these values precisely, but they can be estimated, and the resulting models provide evidence that something like the Snowball Earth episodes *could* happen: glaciers could form even in the tropics, and parts of the ocean could freeze.

An argument against this scenario is that as glaciers spread toward the tropics, even more of the Sun's energy would be reflected and the Earth would simply remain in a deep freeze. Under those conditions, how could the planet ever warm up again?

The answer, or at least *an* answer that fits evidence from the rock record, again focuses attention on the importance of greenhouse gases for the Earth's climate. Joe Kirschvink, who coined the term *Snowball Earth* in 1992, is a geoscientist at the California Institute of Technology who was initially a skeptic about tropical glaciation during the Cryogenian. But when he became convinced, he turned his attention to how a severe ice age might end. He pointed out that even on a frozen Earth, volcanoes would continue to spew out lava containing gases such as carbon dioxide. With low temperatures and most of the continents covered by ice, rock weathering would be minimal; very little carbon dioxide would be used up in the weathering process, and most would remain in the atmosphere. Even photosynthesis, which also removes carbon dioxide from the atmosphere, would be curtailed if the oceans were partly frozen. The volcanic supply of carbon dioxide would greatly outstrip demand, and its concentration in the atmosphere would eventually rise to such high levels that it would counteract Snowball Earth's very high reflectivity. A short-lived "supergreenhouse" climate would ensue and thaw the planet. Other researchers who have embraced Kirschvink's theory have pointed out that the supergreenhouse effect might have been enhanced by the release of methane as the Earth warmed. During very cold periods, much of the methane produced by bacteria is unable to seep out into the atmosphere as it is trapped in solid icy compounds called gas hydrates. As temperatures rose, the methane would have been released in a great burst when the hydrates decomposed (although methane hydrate is icelike, it doesn't actually melt; it just breaks down and releases methane gas).

The severity of the Snowball Earth episodes, and whether or not the following warm periods can really be characterized as "supergreenhouse" intervals, are subjects of intense debate. Some earth scientists prefer the term *Slushball Earth* to imply less extreme glaciation. But extensive field studies over the past few decades, especially those by geoscientist Paul Hoffman and his colleagues, indicate that episodes of extreme cold occurred even in the tropics during the Cryogenian, and that each frigid period was followed by a short interval of high temperatures. Wherever they are still preserved, sedimentary rocks from this period of the Earth's history tell the same story: jumbled layers of glacial deposits are quite abruptly overlain by fine-grained, limestone-like rocks with physical and chemical characteristics indicating they were deposited in warm, shallow seas. The picture that emerges is of an Earth with wild swings in climate, from global deep freeze to global hothouse and back again. Sea level fell dramatically as water was tied up in glaciers, and then quickly rose again as the ice melted.

Because it is difficult to make precise correlations among all of the Earth's Cryogenian rock units, which occur in widely separated localities, there are differences of opinion about just how many ice ages really occurred. Nonetheless, the end of the Cryogenian is generally placed at 635 million years ago, at the well-defined conclusion of a severe period of glaciation. Various chemical indicators in sedimentary rocks show that the oxygen content of the atmosphere was still low at that time, less than 10 percent of present-day values. But less than 100 million years later, at the beginning of the Cambrian period, these same indicators show that oxygen had risen to levels close to those of today. They also indicate that even the deep ocean—until then starved of oxygen—had become oxygenated. Exactly what caused this relatively rapid rise is still unknown, but the chronology of events has led to the suggestion that it may be connected to the climate swings of the Cryogenian and their effect on ocean life. Perhaps the abundance of photosynthetic organisms in ocean surface waters exploded as the climate settled into a more equable mode after the last icehouse-hothouse cycle of the period, facilitated by the large quantities of essential nutrients washed into the seas as rocks ground up by glaciers were weathered. Whatever the cause, however, one consequence was a huge spurt in biological evolution, beginning with the appearance of the first animals in the oceans.

Animals require oxygen to live, so a rise in atmospheric oxygen and oxygenation of the oceans was a prerequisite for their evolution. The oldest known animal fossils come from the Avalon Peninsula of Newfoundland; they date to approximately 575 million years ago. These animals were large, complex, soft-bodied organisms, and they mark the beginning of accelerating eukaryote evolution after more than a billion years of very little change. It is likely that more abundant oxygen was not the only factor in this rapid evolution. These new creatures enjoyed a hospitable climate after the fluctuations of the Cryogenian, and as the glaciers melted back, the seas spread over the edges of the continents, creating a wealth of new ecological niches for them to inhabit.

The 575-million-year-old animals fossilized in the Newfoundland rocks didn't appear instantaneously, of course. They must have had ancestors, but to date no unambiguous physical evidence of those predecessors has been found. There are, however, reports of chemical fossils in rocks slightly older than 635 million years, biomarkers that point to the existence of primitive spongelike animals. There is also a recent report of spongelike fossils in Australian sedimentary rocks of about the same age. If confirmed, these discoveries would mean that some complex animals evolved during, or perhaps even before, the final Cryogenian glaciation. Still, the fossil evidence shows that the main spurt of evolution occurred after the glaciers retreated and the climate had warmed again.

One of the themes that emerges from this brief tour through the Proterozoic eon, from its hazy beginning at the boundary with the Archean to the better-known burst of evolution near its close, is that there are links between geological phenomena that at first don't seem connected in any way. Continent building, greenhouse gases, climate, ocean chemistry, biological evolution—all of these are interrelated at some level. The record in Proterozoic rocks shows just how closely integrated different parts of the Earth system really are.

CHAPTER EIGHT

Cold Times

The ice ages near the end of the Proterozoic eon were undoubtedly the most severe glacial periods ever to affect our planet. But these frigid intervals happened so long ago that we know only their broad outlines and will never be able to fill in many of the fine details. We don't know, for example, the exact configuration of the ice sheets, whether the glaciers waxed and waned repeatedly during the ice ages, how thick the ice was, or the extent of sea level fall and rise in response to glacier formation and melting. But we do know all this and more about a much more recent ice age, one that has influenced humans directly, affecting our landscapes, our climate, and even our evolution. The knowledge accumulated about this recent ice age has given us deep insight into how the Earth's climate system operates, and has made it easier to interpret the clues left in the geological record by ice ages of the distant past. It also provides ground truth for understanding how the climate system may respond to future perturbations.

This recent ice age is usually referred to as the "Pleistocene Ice Age," a name that is slightly confusing. The Pleistocene is a subdivision of the geological timescale that lasted from 1.8 million until 11,400 years ago, yet the Pleistocene Ice Age began earlier, about three million years ago, and has continued right up to the present. On average over this time, the Earth has been much colder than it is today. Certainly, the dominant theme throughout the Pleistocene was glaciation, but there were also warm periods when the climate was at least as mild as it is at present. One of the insights gained from research into the Pleistocene Ice Age is that it is a period of constantly fluctuating temperatures, with long, cold glacial periods, when ice has covered as much as 30 percent of the Earth's surface, interspersed with shorter, much warmer intervals. In the absence of other influences—like the effects of humans—this pattern would be expected to continue, with the current warm interval ending and the Earth cooling again into the next glacial period.

When you come across the term *the last ice age,* it often refers only to the most recent cold glacial interval, not the Pleistocene Ice Age as a whole. In this book I will always try to make the distinction between the ice age—the entire time period—and the glacial and interglacial intervals within it. The Earth began to cool into the most recent glacial interval of the Pleistocene Ice Age about 125,000 years ago, and the coldest point in that interval was about 20,000 years ago. At that time, which geoscientists refer to as the "Last Glacial Maximum" (LGM), thick glaciers in North America extended far south of Chicago.

Nearly all of the evidence concerning glacial advances and retreats during the Pleistocene Ice Age comes from the Northern Hemisphere. Before about three million years ago, as far as we can tell, even Greenland, currently the site of the Northern Hemisphere's most extensive glaciers, was mostly or perhaps even completely free of permanent ice. But geochemical clues from deep-sea sediments indicate that the Earth's average temperature had begun to decrease long before then, and sediment cores from the seas around Antarctica show that large glaciers existed on that continent as early as thirty-five million years ago. By twenty-five million years ago, much of the Antarctic was buried under an ice cap. So in terms of timing, the glacial record is quite different for the two hemispheres, and the obvious question is: Why?

Most earth scientists think it has to do with plate tectonics, in particular the geographical distribution of the continents. About 500 million years ago, today's southern continents were joined together in the large landmass called Gondwana. As this continent broke up hundreds of millions of years later, Antarctica moved to the south pole and the other continents split off and drifted northward. A circumpolar current developed in the southern ocean, completely encircling Antarctica and isolating it from the influence of warmer seas to the north. As glaciers grew on the cold, isolated continent, snow and ice cover caused the amount of solar energy reflected back into space (the albedo) to increase, contributing to further cooling. In contrast, the configuration of continents in the Northern Hemisphere was very different, and ocean currents there continued to transport warm water to high latitudes, delaying the onset of glaciation.

However, when continental-scale ice sheets eventually began to form in the Northern Hemisphere around three million years ago, the Earth slipped into a full-fledged ice age. The realization that this had happened, that large swaths of the continents had been covered by thick glaciers, entered into the general scientific and public consciousness in the 1830s. The idea was developed by the biologist Louis Agassiz, who first documented evidence for a past ice age in his native Switzerland. Agassiz didn't actually make the discovery himself. He was shown the clues—things like ice-carried boulders plopped down in temperate valleys, far from any present-day glacier—by other scientists, who had already concluded that local glaciers must have been more extensive in the past. But it was Agassiz who expanded their ideas into the concept of a global ice age.

Although Agassiz continued his research on the ice age throughout his career, he never really advanced his theory by trying to work out the details of how and where the ice had moved, or why an ice age occurred in the first place. Others did tackle these problems, however. It soon became apparent to geologists mapping out moraines—piles of gravel and boulders scraped up by glaciers and left behind when they recede—that the ice had advanced and retreated many times. Agassiz's ice age was not a single event; rather, it was a series of alternating cold and warm periods. Although this discovery was made long before glacial deposits could be dated accurately, there were enough similarities from place to place for most researchers to conclude that the moraine sequences were everywhere coincident: the glacials and interglacials in Europe, it appeared, had occurred simultaneously with those in North America. That implied that the ice age climate swings were global.

These observations put strict constraints on the possible cause of the ice age. What mechanism could have caused the planet to cool and then warm up again, not just once, but many times in succession? The key, several scientists realized, was the Sun, because the temperature at the Earth's surface ultimately depends on the amount of solar energy received and retained. In 1842 Joseph Adhémar, a French scientist, proposed that variations in the Earth's orbit around the Sun (which changes from almost circular to slightly elliptical), and also in the tilt of its rotation axis, must play a role in glaciation. These two parameters are referred to by astronomers as the "eccentricity" of the orbit and the "obliquity" of the rotation axis, and both affect the amount of solar radiation received by different parts of the Earth's surface. Even in Adhémar's day, it was well known that they change regularly and predictably over long time periods, and he believed the variations were sufficient to affect climate. It eventually turned out that he was on the right track, but the conclusions he reached were so fantastic that his ideas were not taken too seriously (he thought that ice would build up alternately at the north and south poles to such enormous thicknesses that it would shift the Earth's entire center of gravity and produce gigantic tidal waves).

About two decades after Adhémar published his theory, James Croll, a self-taught Scottish scientist and mathematician, extended Adhémar's work by rigorously calculating how the amount of solar energy incident at different latitudes changes as a result of orbital variations. He quickly discovered that the changes are small—too small, he thought, to cause significant variations in global climate. But Croll had other insights. Although he didn't express it in such modern terms, he was an early practitioner of earth system science. Croll realized that climate depends on a complicated interplay of a variety of factors, and that there are multiple possible feedbacks—some reinforcing, some cancel-ing—among the different parts of the climate system. He concluded that if things like heat transfer via ocean currents and changes in the Earth's albedo caused by variations in snow and ice cover were taken into account, the effects of the external astronomical variations might be sufficiently amplified to trigger a glacial episode.

Following the initial excitement created by Croll's work, however, other researchers found flaws in some of his arguments. Also, geologists began to find field evidence that didn't fit his predictions. Remarkably quickly, the "astronomical theory" of ice ages became suspect, and eventually most scientists completely abandoned it. But, almost like the advances and retreats of the glaciers themselves, the idea kept reap-pearing in slightly modified versions. In about 1912 a Serbian engineer turned mathematician, Milutin Milankovitch, embarked on an ambi-tious project: to develop a mathematical theory of the Earth's climate. Milankovitch did all his calculations by hand, computing the average temperature at different latitudes and different times of the year from first principles, taking into account, just as Croll and Adhémar had done, the orbital changes that affect the amount of sunlight—the insolation—received at different points on the Earth's surface. His initial results gave a reasonable match to average observed temperatures around the world, and quickly attracted the attention of meteorologists, who until then had approached climate mainly from an empirical rather than a theoretical point of view.

Several scientists interested in ice ages also took notice of Milan-kovitch's work and urged him to extend his calculations into the past. When he did this, he incorporated changes in the eccentricity of the Earth's orbit and the obliquity of the rotation axis (as Croll and Adhémar had done), but he also included the way the rotation axis, like the wobbling axis of a spinning top, traces out a cone in space (this latter phenomenon, referred to as "precession," is why the point in the night sky that the stars appear to rotate around—today it is the navigator's polestar, or North Star—slowly shifts over time). Eccentricity, obliquity and precession all vary in a regular, well-understood way on timescales ranging from tens of thousands to hundreds of thousands of years. Milankovitch also incorporated details about how the Sun's energy is transmitted through the atmosphere into his calculations. It was a monumental achievement.

Milankovitch's most important result was to show that there are large differences in the way the astronomical variations affect the temperature at different latitudes. He and his collaborators also concluded that glaciation would be most likely at times when summers were cool at high latitudes. This was a crucial point, because others had claimed that cold winters were necessary to initiate glaciation. However, cool summers prevent some of the previous winter's snow from melting, thereby increasing albedo and causing more of the Sun's energy to be reflected back into space. This reinforcing feedback leads to further cooling.

Again there was initial excitement, followed by dissent. There was still the problem that the changes in insolation are so small that on their own they could not possibly control the glacial cycles. There were also problems with timing. Some of the European glacial deposits that appeared to coincide in age with Milankovitch's calculated cool northern summers—and had been used to support his ideas—turned out not to be glacial deposits at all. And when carbon-14 dating was developed in the 1950s, one of the first projects tackled was to measure the ages of glacial deposits in North America. The data showed a much more complex picture than the simple variations in climate predicted by Milankovitch. Once again the astronomical theory of glaciation faded into the background.

Today, however, it is the cornerstone of thinking about the Pleistocene Ice Age, and is usually referred to as the Milankovitch theory, or occasionally the Croll-Milankovitch theory. There is no longer any doubt that the same factors investigated by these men—eccentricity, obliquity, and precession—choreograph changes in climate, although there is still considerable debate about exactly how they do so. For the Pleistocene Ice Age, the alternation between glacial and interglacial intervals seems to be most strongly correlated with insolation variations at a latitude of about $65°$ north.

The breakthrough that confirmed the astronomical theory came from studies of deep-sea sediment cores in the 1970s. These showed that the temperature of seawater in the past, estimated from measurements of temperature-sensitive oxygen isotope ratios in the shells of tiny fossil plankton, has varied in lockstep with the insolation changes calculated from the Croll-Milankovitch theory. The temperature variations show up in sediment cores from around the world, indicating that the changes were global.

Ocean sediment cores have continued to be important in research on ice ages. However, another source of information has revolutionized Pleistocene Ice Age studies in recent decades: deep ice cores from Greenland and Antarctica. The ice cores do not extend as far back into the past as sediment cores, but they do provide a continuous record spanning many of the Pleistocene glacial-interglacial cycles. Crucially, the ice is laid down in distinguishable annual layers, allowing precise dating through many of the cycles. Trapped in the ice are clues to the temperature, the composition of the atmosphere, and even the strength of the wind during a particular time period, as well as details of volcanic eruptions. This information has proved invaluable for testing theoretical climate models, because only simulations that are able to reproduce tens of thousands of years of real climate variations recorded in the ice cores are likely to be reliable for predicting climate in the future.

The realization that glacier ice might be useful for peering into the past goes back at least to the 1930s, when a German scientist, Ernst Sorge, took part in a meteorological expedition in Greenland. Conditions were rigorous; he spent the winter holed up in a snow cave for protection against the elements. But when weather permitted, he left his den and laboriously dug a deep pit nearby. He carefully examined and documented the layers in the wall of this pit, and understood that they held an annual record of conditions in central Greenland extending back to some indeterminate time in the past. Modern ice core drilling to exploit this record only began in the 1950s, however, primarily the result of work done by the U.S. Army Corps of Engineers. It was the time of the Cold War, and the Army wanted to know if they could build an airfield and land heavy aircraft on the glaciers of Greenland. But first they had to find out as much as they could about the physical properties of the ice and snow. In a typically symbiotic relationship, many of the scientists who became involved in the research were interested in the purely scientific aspects of glacier studies, such as how surface snow transformed into solid ice, how the ice changed with depth, and especially whether they could retrieve a record of past conditions from ice cores.

The first truly deep ice drilling occurred on Greenland in 1966 at a U.S. base called Camp Century. The gigantic operation, taking almost six years to complete, retrieved the first ever continuous ice core through a glacier from top to bottom, eventually reaching bedrock at 4,550 feet, or more than eight-tenths of a mile, below the surface. The drill team had to circumvent the problems of drilling through constantly moving ice (continuous flow occurs even deep within a massive ice sheet) and the tendency of the drill hole to collapse because of the pressure of the surrounding ice. The work was done under difficult conditions; the Greenland icecap is not especially hospitable at any time of year, let alone in the dead of winter. But, groundbreaking as the work at Camp Century was, the drilling team didn't rest on their laurels. Just a few months after completing their work in Greenland, they hauled the drilling equipment to the other end of the Earth and set it up again at Byrd Station in the Antarctic. A year and a half later—with the experience of Greenland drilling under their belts—they reached another first: a continuous core through the Antarctic ice sheet, this one more than one and one-third miles deep.

During the drilling at Camp Century, sections of the ice core were examined and photographed as they were retrieved, various measurements were taken, and the ice was stored at low temperatures for future

study. Samples were also sent to a scientist in Copenhagen, Willi Dansgaard (who wasn't allowed to visit the drilling site and select his own
samples as it was within a restricted U.S. military zone). Dansgaard had
earlier discovered that oxygen isotopes in rain and snow are indicators
of temperature and had predicted that the annual layers of Greenland
ice might preserve a record of past temperatures. He initially thought
the Camp Century core might probe back several hundred years, but it
turned out to cover more than *one hundred thousand* years of history. His
measurements—the first of thousands of oxygen isotope measurements
to be carried out on polar ice cores   revealed in detail how Greenland
temperatures had fluctuated over that period (see figure 22).

Since those first analyses, oxygen isotopes have been a cornerstone
of ice core research. Other methods have also been developed to track
temperature, and they confirm the oxygen isotope results. Greenland
ice provides a record of only one of the Pleistocene Ice Age's multiple
glacial-interglacial cycles, but cores from the Antarctic extend through
many, and the fine-scale time resolution available from the ice layers
means that climate change can be examined in exquisite detail through
the cooling and warming intervals. The availability of multiple cores
from both Greenland and the Antarctic also means that temperature
patterns retrieved from different drilling sites can be compared to ensure
that the chronology of each core is interpreted correctly. Long-distance
comparisons, between Greenland and the Antarctic, have proved especially revealing; while they show that temperature changes during the
glacial cycles were global and roughly coincident, they also indicate that
they were often of different intensity on different parts of the Earth—a
reflection of complexities in the climate system that are still only partly
understood.

Important as the temperature information is, the light that ice cores
shed on greenhouse gases in the atmosphere is equally valuable. But
how do scientists retrieve information about greenhouse gases from
glacial ice? The answer is actually quite simple. As the snow that falls
on glaciers gets compressed and turned into solid ice, small air bubbles

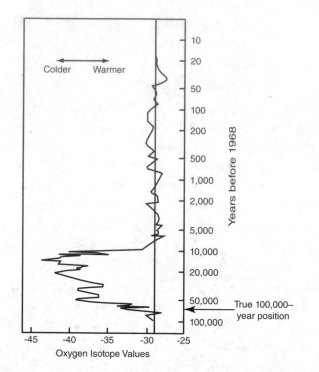

Figure 22. Willi Dansgaard's original oxygen isotope data for the ice core retrieved at Camp Century, Greenland, in 1966. The time axis is logarithmic and shows time in years before 1968. The timescale Dansgaard used here was based on a simple model for ice accumulation and is now known to be slightly inaccurate beyond about 15,000 years; the true location of the 100,000-year level, based on more recent calibrations, is indicated along the vertical axis. As noted in the diagram, smaller oxygen isotope values correspond to colder temperatures, while higher values indicate warmer temperatures; the solid vertical line corresponds to today's conditions. (Based on figure 6–1 in Dansgaard 2000; copyright Centre for Ice and Climate, Niels Bohr Institute, Copenhagen.)

are trapped. The upper layers of ice cores from Greenland and the Antarctic are full of visible bubbles, tiny time capsules that contain samples of the atmosphere as it was when the ice formed. Deeper in the cores, the bubbles become smaller and eventually disappear due to the great pressure, but the air can still be recovered simply by melting the ice. The air samples obtained in this way are very small, but modern analytical instruments are capable of measuring even the minute amounts of trace gases, such as carbon dioxide and methane, that they contain.

In the interior of Antarctica, temperatures remained low enough to prevent glaciers from melting even during interglacial periods, which is why ice cores from that continent extend much farther back into the past than those from Greenland. Figure 23 shows data from one of these cores, and illustrates how temperature and other properties can be tracked in unprecedented detail. Several observations can be made from such data. First, although there is a periodic alternation in temperatures, the changes are not perfectly regular. On casual inspection, the peaks (warm periods) in the temperature panel of figure 23 look as though they occur quite regularly every 100,000 years, but a closer look shows that they have variable durations (this is even more obvious in the few Antarctic records that extend even further back in time). In addition, the shapes of the peaks and troughs differ from one interval to the next. Second, temperature increases at the beginning of the warm intervals are much more rapid than the cooling that follows them, giving the graph a sawtooth appearance. And finally, the warm periods are very short. Over the past million years or more, average temperatures have been much colder than they are today.

What can such observations tell us about the Pleistocene Ice Age cycles in particular, and in general about how the Earth's climate system responds to various forcing factors (climate scientists use the term "forcing factors" to describe almost anything that acts to change climate, from greenhouse gases to albedo and astronomical parameters)? The most obvious feature of the ice core data is the 100,000-year temperature

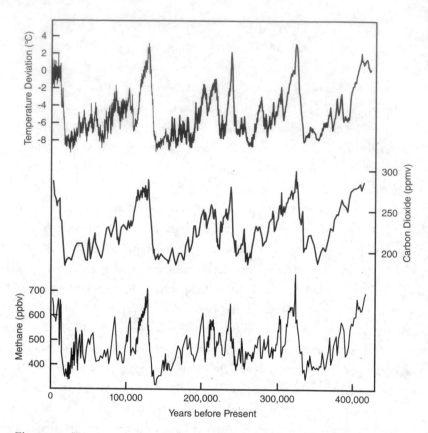

Figure 23. Temperature, atmospheric carbon dioxide, and atmospheric methane over the past 400,000 years, as measured in the Vostok ice core from Antarctica. Note that the temperature is given as the deviation from the present day, not the absolute temperature, and is in degrees Celsius (one degree Celsius = 1.8 degrees Fahrenheit). Carbon dioxide concentrations are in parts per million by volume; the present-day concentration shoots off the scale at 385 ppmv and is not shown here. Methane concentrations are in parts per billion by volume. (Data from J. R. Petit et al., *Vostok Ice Core Data for 420,000 Years*, IGBP PAGES/World Data Center for Paleoclimatology Data Contribution Series 2001–076 [Boulder, CO: NOAA/NGDC Paleoclimatology Program, 2001.] See also Petit et al. 1999.)

cycles, and this fits neatly with Milankovitch's theory, because the eccentricity of the Earth's orbit around the Sun changes on the same timescale, from almost circular to slightly elliptical and back again over 100,000 years. The two additional factors that Milankovitch considered, obliquity and precession, have quite different timescales; they operate on cycles that last approximately 41,000 and 23,000 years, respectively. These time periods are not immediately obvious in graphs like figure 23, but it is the interaction among all three of the astronomical parameters that ultimately determines how much solar energy reaches various parts of the Earth. If graphs like figure 23 are separated into their component parts using a technique known as spectral analysis, it turns out that their overall shape is indeed the combined result of changes on 100,000-, 41,000-, and 23,000-year timescales. Spectral analysis of this kind is a bit like taking the sound of a symphony orchestra and separating out all of the individual musical frequencies that make up the final sound; some frequencies are more important than others. In the case of the temperature record, spectral analysis shows that the most important frequencies are those of the three astronomical cycles, providing convincing evidence that these external factors really do have an important influence on the Earth's climate. Because the astronomical cycles have different timescales, they sometimes reinforce and sometimes interfere with or cancel out each other's effects; the net result is glacial-interglacial fluctuations that are not perfectly regular.

Why the temperature record has a sawtooth appearance, with slow cooling into glacial periods but very abrupt transitions to the warmer interglacial intervals, is less clear. It seems likely that warming occurs when the astronomical parameters all align to simultaneously increase the amount of solar energy received in both the Northern and the Southern Hemispheres above some threshold level. But why the change is so rapid is still obscure. Almost certainly, the insolation changes are amplified by factors such as albedo, greenhouse gases, and changes in ocean circulation.

Work on the glacial-interglacial cycles of the Pleistocene Ice Age has highlighted the importance of such amplification or "feedback" factors for the Earth's climate system. It has also shown that interactions among these mechanisms are exceedingly complex; small changes in one can sometimes cause unexpectedly large variations in another. However, climate researchers have gained a good general understanding of how the most important feedback factors operate and of some of the interactions among them.

It turns out that albedo is one of the most important. It has a direct effect on average temperature; high albedo is so effective at cooling the Earth that if our planet were covered with a white blanket, it would freeze solid (this is why only a "supergreenhouse" can explain how the Earth recovered from the Proterozoic Snowball Earth episodes). The nature and distribution of different types of surfaces are especially important for regulating the Earth's overall albedo. Oceans, for example, reflect much less of the Sun's energy than do continents, and therefore oceans absorb more heat. Deserts absorb less solar energy than do rain forests, and snow and ice absorb almost none. It has been suggested that a cost-effective way to combat global warming would be to make all roofing material white, thereby increasing the Earth's average albedo. The dramatic amplifying effect of relatively small changes in albedo has been illustrated in the past few years by the rapid warming of the Arctic. Until fairly recently, sea ice that melted in the summer warmth was replaced with new ice during the winter freeze, keeping the Arctic's albedo roughly constant over the annual cycle. But as the Earth's climate has warmed, formation of new winter ice has not kept pace with summer melting, and an increasing amount of open ocean is being exposed to the Sun's rays each year. In a quickly escalating sequence, that has led to accelerated warming, rapid melting of glaciers in Greenland, the opening up of northern shipping routes that until recently were considered impassable because of ice, and a loss of habitat for polar bears. Although the importance of albedo has long been known, the speed and intensity of Arctic warming has caught most earth scientists by surprise, with temperatures increasing much faster than anyone imagined even a few years ago. Some scientists predict that within just a few decades the Arctic Ocean will be entirely ice-free during the summer months.

Greenhouse gases also play an important feedback role in the climate system. Data from ice cores show that there is a direct correlation between temperature and the amount of greenhouse gas in the atmosphere, with smaller concentrations during cold glacial periods and higher concentrations during the warm interglacials (figure 23). When this correlation was first discovered, it was unclear which was cause and which effect. But careful examination of the chronology of these changes shows that carbon dioxide content lags temperature by at least a few hundred years; changes in carbon dioxide levels must therefore be a response to glacial-interglacial climate changes, not a cause. A similar situation exists for methane.

Higher methane concentrations during warm periods are easy to understand, because a major natural source of methane is bacterial decomposition of organic material in bogs and wetlands, and warm temperatures enhance this process. Also, when high-latitude permafrost thaws during the warm interglacials, boggy ground spreads, further enhancing methane production. During glacial periods, these processes are reversed.

For carbon dioxide, however, the picture is less clear, at least partly because the sources for atmospheric carbon dioxide are more complex than those for methane. On the timescale of the glacial-interglacial changes, its concentration in the atmosphere depends on several factors, including biological cycling of carbon and the temperature and circulation pattern of the oceans. The oceans are especially important because they hold a very large amount of dissolved carbon dioxide (today almost half the carbon dioxide emitted into the atmosphere by humans is soaked up by the oceans and will be stored there for thousands of years or more). But regardless of the processes involved, the rise in carbon dioxide concentration during interglacials acted as a powerful reinforcing feedback that amplified warming. Decreases during glacial intervals similarly amplified cooling, and the changes in methane concentration had a similar effect. Simulations of Pleistocene Ice Age temperature changes simply don't match the ice core records unless the feedback effects of changing greenhouse gas concentrations—particularly for carbon dioxide—are taken into account.

There are other feedback factors, too, although for the most part they are not as well understood as albedo and greenhouse gas variations. They include several phenomena that operate indirectly by changing the albedo. As glaciers grow, for example, sea level falls, increasing the total area of the continents and thus increasing the Earth's overall albedo. Large changes in vegetation cover through the glacial-interglacial cycles also affect albedo. As temperatures fall, vegetation cover decreases markedly, especially at high latitudes and altitudes. Albedo increases, and so does the amount of dust in the atmosphere (an effect that is recorded by an increase in the amount of dust in ice cores). A dusty atmosphere partially blocks incoming sunlight, providing another reinforcing feedback. During the warm interglacials, higher temperatures enhance evaporation, increasing humidity—and because water vapor is a powerful greenhouse gas, this pushes temperatures even higher. Each of these additional feedback factors acts to amplify the temperature trends of the ice age cycles.

The overall picture that has emerged from many decades of research on ice cores, ocean sediment cores, and other climate records is that Milankovitch's astronomical variations have triggered and paced the swings between glacial and interglacial climates during the Pleistocene Ice Age, but that the internal responses of the Earth's climate system have then taken over, amplifying the trends. High latitudes have the largest insolation changes during the Milankovitch cycles, and have also experienced the largest temperature variations. The Northern Hemisphere seems to be especially important in choreographing the ice age cycles, perhaps because of the configuration of the continents, with the Arctic Ocean at the pole and continents surrounding it.

With an understanding of how past glacial cycles have worked, it is

possible to look into the future and ask when the astronomical parameters will next align to give high northern latitudes exceptionally cool
summers, the condition that appears to have triggered past glacial
periods. Warm interglacials have usually been short, ten thousand to
twenty thousand years long, so purely from a perusal of graphs such
as figure 23 you might think a new ice advance is imminent. Indeed,
when the chronology of glacials and interglacials was first worked out,
many scary stories appeared in the media about the probable return
of the glaciers. But you needn't worry, nor should your grandchildren.
The orbital parameters have conspired to make the current interglacial period an especially long one; calculations of future changes in
Northern Hemisphere insolation indicate that there is little chance of a
glacial advance for at least another thirty thousand years. And that prediction doesn't take into account enhanced warming due to fossil fuel
burning, which will probably reduce high latitude ice to a greater extent
than has occurred during past interglacials. Because of the albedo feedback, that would further extend the warm period.

We may not have to be concerned about the return of a glacial climate any time soon, but the high-resolution record from ice core data
has given scientists something else to worry about: very abrupt changes
in climate. The ice cores show that at times in the past, temperatures
in Greenland changed drastically on timescales of decades or less. How
the Greenland temperature swings translate to other parts of the world
is not known with certainty, but even much smaller changes, if they
occurred on such a short timescale, would have devastating effects on
agriculture and human societies worldwide. Many earth scientists are
now working on abrupt climate change, but its causes are still obscure,
and prediction is problematic.

The ice cores show that over about the past eleven thousand years—
roughly the time since the Northern Hemisphere emerged from its
last glacial period—the temperature has been remarkably stable (see
figure 24). There is one significant blip, at about 8,200 years ago, when
there was a short-lived cold period (less than two hundred years long);

but through the rest of the period, temperatures have fluctuated more or less randomly over a small range. By comparison, temperature changes in the earlier part of the record are huge. A large drop about thirteen thousand years ago ushered in a cold period known to climatologists as the "Younger Dryas event" (named after a plant, normally found in the Arctic, that flourished in Europe during this interval). The Younger Dryas lasted about a thousand years and ended when the temperature shot up by an enormous eighteen degrees Fahrenheit in a decade. Even farther back in time, the ice cores reveal a series of abrupt, brief warm intervals that interrupted the glacial chill (figure 24). These are referred to as "Dansgaard-Oeschger events," after the two scientists who first discovered them. During these short temperature excursions, Greenland warmed to near-interglacial conditions in a matter of decades, and then, slightly more slowly, temperatures decreased again to their earlier levels. In spite of their short duration, the Dansgaard-Oeschger events recognized in the Greenland ice cores seem to have been global in nature. Their effects can also be seen in various paleoclimate indicators from as far afield as China and the Caribbean.

These short-lived climate changes were too rapid and occurred too frequently to be associated with astronomically caused insolation variations. Just as for the glacial-interglacial fluctuations, bubbles in ice cores show that they were accompanied by increases in greenhouse gas concentrations during warm periods and decreases during cold intervals. Also mirroring the much longer glacial-interglacial variations, the greenhouse gas changes followed rather than led the temperature changes, probably for the same reasons. Most paleoclimatologists have concluded that these rapid events resulted from small perturbations in the Earth's climate system that pushed it abruptly from one relatively stable state into another. Exactly what these perturbations were is unknown, but they probably involved changes in ocean and atmospheric circulation, which are the primary mechanisms for heat transport around the globe.

An important collaborating clue for this conclusion is the fact that the abrupt changes were more extreme in Greenland than in the Ant-

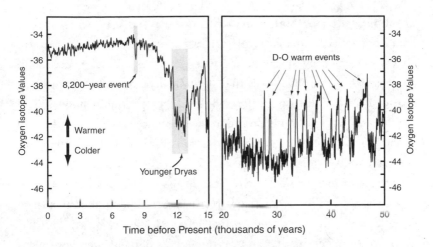

Figure 24. *Left:* The oxygen isotope data for an ice core from northern Greenland shows that the climate has been very stable over the past 8,000 years compared to earlier times. Shading highlights two cold intervals that started and ended abruptly: the Younger Dryas, which started around 13,000 years ago, and a smaller event 8,200 years ago. *Right:* Oxygen isotope data for an earlier section of the same ice core show numerous Dansgaard-Oeschger (D-O) warm intervals between 20,000 and 50,000 years ago, when the Earth was in the grip of a cold glacial period. Note that the timescale is different for the two panels. (Based on data from the North Greenland Ice Core Project [NGRIP] team, measured at the Niels Bohr Institute, University of Copenhagen; see Andersen et al. 2004.)

arctic. Furthermore, many of the larger changes were out of phase between the two hemispheres. Antarctic ice cores, for example, show warming just as Greenland temperatures dropped during the Younger Dryas event. A plausible but still unproven explanation is that the flow of warm surface waters from the tropics into the North Atlantic region, accomplished today by the Gulf Stream, slowed or stopped, cooling Greenland but leaving more warm water than usual in the Southern Hemisphere, keeping that hemisphere warmer as a result.

But what would cause such a change in ocean circulation? For cold periods like the Younger Dryas and the event at 8,200 years ago, there are good candidates. Both intervals occurred as the world was in transition from a glacial to an interglacial period. Northern Hemisphere ice sheets were melting and releasing vast quantities of fresh water; often this was stored in huge lakes that periodically broke through a melting ice barrier and emptied into the ocean in enormous floods. Field studies and careful dating have identified occurrences of this kind both at the beginning of the Younger Dryas and just before the abrupt cooling 8,200 years ago. In the case of the Younger Dryas, a huge glacial lake that covered parts of central Canada and the United States flooded northwest along the edge of the retreating icecap and up the Mackenzie River drainage system to the Arctic Ocean. By 8,200 years ago, the glaciers had retreated further and vast quantities of meltwater were again dammed up against their southern margins. Once again an ice barrier collapsed, sending a gigantic flood of fresh water northward, this time directly into Hudson Bay and the North Atlantic.

Today salty water flows north from the tropics into the North Atlantic, cools and becomes very dense, sinks, and forms a deep water current heading south. Like a conveyor belt, this process continually brings warm water into the North Atlantic and exports cooler water back at depth. But when large quantities of low-density fresh water flooded into the Arctic Ocean and the North Atlantic, the density of surface water decreased, preventing it from sinking. The conveyor belt stopped, or at least slowed substantially. Heat normally transported northward remained in the tropics; the North Atlantic region cooled and the Antarctic warmed. This scenario can explain abrupt cooling in Greenland 8,200 years ago and during the Younger Dryas interval, but the cause of the warm Dansgaard-Oeschger intervals is less obvious. They too are out of phase between Greenland and the Antarctic, and must have involved changes in the circulation patterns that transported heat between the two hemispheres. The trigger for these circulation changes, however, is obscure.

The field of abrupt climate change is still young. Nearly every new issue of the important scientific journals, it seems, carries new data about rapid climate change gleaned from glacial ice or lake sediments or cave stalactites—anything that accumulates slowly and can be analyzed for clues to past climates. Ice core drilling has been extended from polar regions to small glaciers in the Andes and Himalayas to gain a more global view of past temperature changes, and new methods of analysis are constantly being developed. To some extent all of these efforts are driven by scientific curiosity about the Earth, but they have also been given a huge boost—and a sense of urgency—by widespread concern about how climate will evolve in the future. The greenhouse gases already injected into the atmosphere from fossil fuel usage will warm the planet for many centuries, and the temperature rise will accelerate if emissions are not slowed soon. The goal for many climatologists is to determine at what point such changes might tip the Earth into a different climate state that would end the relative climate stability the world has enjoyed for many thousands of years and initiate the kinds of abrupt climate fluctuations that characterized earlier times. So far the goal has proved elusive. But we should all hope that the climatologists are successful.

CHAPTER NINE

The Great Warming

In 1991 a paper in the journal *Nature* reported what the authors called "a remarkable oxygen and carbon isotope excursion . . . near the end of the Paleocene." To anyone not familiar with the intricacies of paleoclimatology, this might sound like so much gobbledygook. You might also wonder why anyone should care. But the "isotope excursion" described in this paper, which was measured in deep-sea sediments dating from about 55 million years ago, signaled that global temperatures had increased dramatically and very suddenly at that time—and also that ocean circulation had undergone a radical reorganization. Not only that, the changes coincided with the extinction of many species of foraminifera, a common variety of ocean plankton. Clearly, something unusual had happened at the Earth's surface. But what?

The authors of the *Nature* paper, James Kennett from the University of California at Santa Barbara and Lowell Stott from the University of Southern California, were studying the 55-million-year-old sediments because they wanted to find out what had caused the sudden demise of so many species of plankton. They could not have imagined the interest their research would generate. Even now, two decades after their publication, the rapid climate change they documented is a frequent subject of talks at scientific meetings and papers in scientific journals, and the pace of research is, if anything, accelerating. The interval identified by Kennett and Stott is often referred to as "The Great Warming." More formally, it is known as the Paleocene-Eocene Thermal Maximum (PETM).

The PETM occurred at a time when the Earth as a whole was much warmer than it is now; even the polar regions were largely free of permanent glaciers. Information about climate from that time comes primarily from isotope studies of deep-sea sediment cores, especially oxygen isotope ratios in the shells of fossil plankton. These ratios reflect the water temperature when the organisms were alive, a good indicator of the general climate. The isotope data show that even during what most geoscientists thought was a period of mild, stable climate, there were occasional short periods when global temperatures increased rapidly, remained high for a short time, then decreased again to near their previous values. The PETM was among the most extreme of these.

But why all the excitement about a short interval of warm temperatures that occurred tens of millions of years ago? Partly, it has to do with the unusual nature of the PETM; anomalous events in the geological record always attract attention because they may provide especially useful insight into how the Earth system works. But also it has come to be realized that the isotope excursion described by Kennett and Stott was caused by the injection of a very large amount of carbon (in the form of the two carbon-containing compounds, carbon dioxide and methane) into the ocean-atmosphere system over a short period of time. All available evidence indicates that the amount added was comparable to, or perhaps exceeded, the quantities that will be released over the next few centuries as a result of human activities. That may make the PETM a good analog for anticipating future climate change.

In spite of all the work being done on the PETM, however, it is still often referred to as a "mystery." Where did all the carbon dioxide come from, and what triggered its release? And where did it all go in the end? In a sense the geologists and geochemists examining these and other questions about the PETM are like detectives, assembling the available evidence and forming hypotheses about how it might have happened. Sometimes one hypothesis or another can be scratched off the list because it doesn't fit with newly discovered evidence; sometimes a new hypothesis is added when additional clues are uncovered. So far, however, no hypothesis has been universally accepted as the "right" one; the mystery, although a bit less mysterious than it used to be, is not yet completely solved.

Let's examine some of the evidence and see what it implies about possible causes. One of the most striking characteristics of the PETM is the large, sharp change in the ratio of two carbon isotopes, carbon-13 and carbon-12; this is the carbon isotope excursion discovered by Kennett and Stott. Subsequent work has shown that the isotope shift did not occur only in plankton shells, but also permeated carbon in all its diverse forms, on land, in the oceans, and in the atmosphere. That got scientists very excited, because such a large and pervasive change indicates that there was a profound disturbance to the Earth's "carbon cycle."

Geochemists have long realized that for many of the chemical elements there are strong interconnections between what happens in different parts of the Earth: processes in the crust can affect the ocean, processes in the ocean can affect the atmosphere, and so on. Conceptually, a convenient way to think about these interconnections is in terms of geochemical cycles: many of the chemical elements cycle through different parts of the Earth and, eventually, return to their starting point. In the Earth's surface environment, there are often feedback processes that keep the concentrations of these elements in balance, or at least within certain limits, preventing them from piling up in one part of the Earth at the expense of another.

Carbon has an especially interesting (and complex) geochemical cycle. It is a crucial component of all living things; every plant, animal, and bacteria cell contains carbon, and the carbon cycle is therefore affected by biological as well as inorganic processes. Also, two of the most important greenhouse gases, carbon dioxide and methane, are carbon-bearing compounds and important players in the carbon cycle. In various forms, huge amounts of carbon continually move among different "reservoirs" on the Earth: the ocean, the atmosphere, the bio-sphere (all living things), soils, sediments, and lakes. Photosynthesis is an important part of the cycle; plants take carbon dioxide from the atmosphere and use it to make the organic carbon of their cells, releas-ing oxygen back into the atmosphere in the process. When the plants die, the organic carbon may be oxidized and quickly returned to the atmosphere as carbon dioxide, or it may be stored in soil or sediments for thousands or even millions of years. In bogs and other environments where very little oxygen is available, bacteria transform the carbon of dead plants into methane, which bubbles up in eruptions of "swamp gas" and escapes into the atmosphere. Sometimes the organic carbon evades attack by bacteria and forms beds of peat, or, if the peat is buried deeply and heated, coal. Carbon stored as peat or coal is removed from the active carbon cycle for very long periods of time.

Photosynthesizing plankton living in the sunlit, near-surface waters of the ocean use carbon dioxide to make organic carbon just as land plants do, and some of them also extract dissolved carbon from sea-water to make shells of calcium carbonate. When these organisms die, both the shells and the organic remains sink to great depths, effectively removing carbon from the ocean surface layers. Some of this carbon redissolves in the deep water, and some is deposited on the seafloor and stored as ocean-bottom sediment. Geochemists refer to this process as a "biological pump," because as it depletes surface water of carbon diox-ide and sends carbon to the ocean depths, it paves the way for additional atmospheric carbon dioxide to dissolve in the now-depleted surface water. The concentration of greenhouse gases in the atmosphere is thus in a delicate balance, affected by the various processes involved in the natural carbon cycle. The balance, at least over short timescales, can be disrupted by many different kinds of environmental change, including independently caused temperature and climate changes. For example, because cold seawater holds more dissolved carbon dioxide than warm water, an increase in seawater temperature causes some of the ocean's vast store of the gas to be released, and its concentration in the atmo-

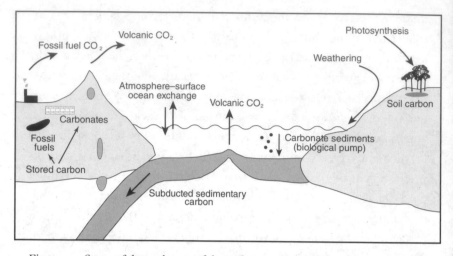

Figure 25. Some of the pathways of the carbon cycle. Different parts of the cycle operate on quite different timescales. Over millions of years, carbon (in the form of carbon dioxide) is taken out of the atmosphere by rock weathering and photosynthesis, deposited as carbon-containing sediments in the oceans or organic matter in bogs and swamps on the continents, and stored for long periods as carbonate rocks or fossil fuels. This is balanced by carbon dioxide added to the atmosphere through volcanism. Carbon also cycles through the atmosphere, plants, soils, and the oceans on shorter timescales.

sphere rises; conversely, a decrease in seawater temperature causes more atmospheric carbon dioxide to dissolve in the ocean. Figure 25 illustrates both short- and long-term pathways of the carbon cycle.

There is an additional important twist to the carbon cycle. As described briefly in chapter 7, over the very long term—millions of years and longer—carbon dioxide in the atmosphere is regulated by input from volcanic activity on the one hand, and by chemical weathering of surface rocks on the other. Lava erupting at the Earth's surface brings with it dissolved gases from the interior, including carbon dioxide. If the volcanic supply is high, carbon dioxide builds up in the atmosphere and temperatures increase due to the greenhouse effect. But as its concentration increases, more carbon dioxide dissolves in rainwater, making the rainwater acidic and increasing the intensity of chemical weathering. The higher temperatures also accelerate the weathering process. This acts as a negative feedback; the net effect of higher atmospheric carbon dioxide is to stimulate chemical weathering, bringing the carbon dioxide concentration down again.

Chemical weathering of surface rocks produces dissolved, carbon-bearing compounds that are carried to the ocean by rivers and eventually deposited on the seafloor as carbonate rocks like limestone. If the limestone is transported down a subduction zone into the Earth's interior and heated up, some of the carbon may be returned to the atmosphere as volcanic carbon dioxide, completing the cycle. Limestone that escapes this fate may sequester carbon that was once in the atmosphere for many hundreds of millions of years.

When the volcanic supply of carbon dioxide is low, the balancing forces work in reverse—weathering slows down, less carbon dioxide is utilized, and the gas builds up in the atmosphere. In this way carbon dioxide concentrations are maintained within broad limits that keep the Earth's surface habitable, neither a frozen wasteland (as it would be without greenhouse gases) nor a broiling inferno like the surface of Venus, which has an atmosphere made up largely of carbon dioxide. Of course, this is a somewhat simplified picture; other factors are involved too. But the general concept of a balance between volcanic supply on the one hand and consumption by weathering on the other is useful for considering the long-term controls on atmospheric carbon dioxide.

The geological record from the PETM, however, indicates that short-term processes completely disrupted the carbon cycle's long-term balancing act. The change in carbon isotope ratios recorded in deep-sea sediments at the beginning of the interval occurred very abruptly, within a few thousand years (see figure 26). The rapid start implies that a large amount of carbon was suddenly injected into the ocean-atmosphere system, and the unusually large change in the carbon isotope ratios indicates that the injected carbon had a very different isotope ratio from "normal" carbon. That quickly pointed the finger of suspicion at methane.

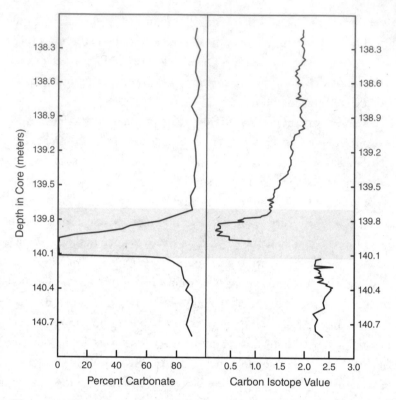

Figure 26. Data through the Paleocene-Eocene Thermal Maximum (PETM) from a sediment core recovered in the South Atlantic. In this core, the PETM begins at a depth just below 140.1 meters, where the carbonate concentration drops rapidly to zero; the concentration recovers to its pre-PETM values after about 110,000 years. Note that carbon isotope data are missing through the first part of the PETM—there are simply not enough carbonate shells for analysis in this interval. The section of the core over which no carbonate was deposited corresponds to about 60,000 years. (Data from Zachos et al. 2005)

Why methane? In the units used to report the isotope ratio between carbon-13 and carbon-12, sediments from immediately before the PETM have values near +2.5. Shortly after the start of the warm period, the value dropped to near zero, a decrease of almost three units. Similar changes occurred in organic carbon on land. For the carbon isotope world, that is an enormous shift. The total amount of carbon cycling through the various reservoirs in the oceans, the atmosphere, and on land is so large that to change its isotope ratio by even one unit, let alone three, requires a massive input of carbon. The closer the isotope ratio of the added carbon is to that already in the system, the more carbon is required. Biologically produced methane has an extreme carbon isotope ratio—the value is about -60. Even with this extreme ratio, however, the amount added would have to be in the range of 2,000 to 4,000 *billion* tons to produce the observed isotope change. If the carbon came from any other source, even more would be necessary.

Normally, methane produced by bacterial decomposition of organic matter slowly seeps out into the atmosphere, or, if the bacteria live in ocean sediments, into seawater. That is what happens today in swamps, on the seafloor, and even in garbage dumps and sewage plants. However, something very different from this slow, steady leakage must have occurred to initiate the PETM. One of the first suggestions was that methane hydrate was involved, the same compound some scientists think may have contributed to the "supergreenhouse" climate intervals that followed the Snowball Earth episodes of the Proterozoic eon. Methane hydrate is a solid, icelike substance that forms only at low temperatures in special environments such as the Arctic tundra, or in certain types of ocean sediments. Exploration of the continental shelves, the relatively shallow parts of the ocean surrounding the continents, has shown that layers of this compound are abundant in the sediments of these regions today. Methane hydrate holds so much methane—about 170 times more than the amount contained in an equivalent volume of methane gas—that these deposits are being actively investigated as an alternative source of natural gas (it is estimated that the hydrate deposits hold more methane than all conventional sources of natural gas). It is a strange substance—when dredged up from the seafloor, methane hydrate looks just like a chunk of ice, but touch a match to it and it bursts into flame.

Methane hydrate is only stable over a limited range of temperatures and pressures, which is why its occurrence is restricted to specific environments. The deposits of the continental shelves form when methane produced by sediment-dwelling bacteria gets trapped in icy layers due to the low sea-bottom temperatures and the high pressure imposed by overlying water and sediments. It has the necessary extreme carbon isotope ratio because of its biological origin, but whether there was enough methane hydrate on the continental shelves 55 million years ago to account for the PETM carbon isotope shift is still an open question. There is also another problem. The abrupt isotope ratio change at the beginning of the PETM would have required rapid release of an amount of methane hydrate equivalent to the total in existence today. What could have triggered such an extraordinary event?

Perhaps you are beginning to see why I earlier compared geochemists investigating the PETM to detectives. Each clue seems to point to a suspect, but further probing raises many questions. And clues that are 55 million years old are sometimes hard to decipher. In some ways it is remarkable that we know as much as we do.

There are several mechanisms that could have caused the abrupt methane release. One is an increase in the temperature of deep ocean water, leading to decomposition of the methane hydrate layers on the continental shelf. But why might the bottom water have heated up? If climate change was the cause, then what might have brought about this change? It is also possible that a shift in ocean circulation—without an overall increase in seawater temperature—simply brought warmer water to some regions, triggering methane release, which in turn caused a general warming of the climate, which triggered more methane release in a kind of snowball effect. Again there is the question of what could have caused the initial circulation change. Another possibility is that landslides along the edges of the continental shelves destabilized hydrate layers, releasing the methane. This too has its difficulties—the process would have to have occurred globally over a short period, which seems unlikely. It is also quite possible that some combination of different mechanisms acting more or less simultaneously was responsible. Regardless of the release mechanism, however, the incontrovertible fact is that the large size of the carbon isotope shift seems to require that much of the added carbon was in the form of biologically produced methane.

On a volume-for-volume basis, methane is about eight times more effective than carbon dioxide at trapping heat in the atmosphere, so the injection of huge quantities of the gas would cause a rapid increase in global temperature. However, the average residence time of a methane molecule in the atmosphere before it is destroyed by oxidation is less than ten years. This is far too short to account for the entire period of PETM warming, which, although it began abruptly, continued over thousands of years until the peak temperatures of the interval were reached. But an initial rapid release of truly massive amounts of methane would have temporarily overwhelmed the atmosphere's oxidizing capacity, lengthening methane's atmospheric life span many times over. Under such circumstances, the enhanced greenhouse effect and rising temperatures could have persisted for centuries, spawning additional release through further melting of hydrates. Land-based production of methane by bacteria in swamps and wetlands would also have increased as the climate warmed.

Furthermore, when methane is oxidized, carbon dioxide is produced. Carbon dioxide has a much longer atmospheric residence than does methane, so its concentration would have built up as methane was oxidized, sustaining the initial warming. This conclusion is corroborated by unequivocal evidence from deep-sea sediment cores that the PETM atmosphere had greatly elevated levels of carbon dioxide. In the cores, the beginning of the PETM—the point where carbon isotope values drop rapidly—is marked by a layer of clay. This is not due to an increased supply of clay particles to the seafloor; it is the result of a *decrease* in the supply of other kinds of particles, particularly the calcium carbonate shells of planktonic organisms such as foraminifera, which are normally a predominant component of ocean sediments. The change in calcium carbonate content is apparent in ocean sediments from around the globe, and is one of the most striking characteristics of the PETM (see figure 26).

But how is this feature related to carbon dioxide in the atmosphere? The precipitous drop in calcium carbonate—from nearly 100 percent of all sedimentary particles to virtually zero—occurred because the ocean became more acidic, causing sinking plankton shells to dissolve before they reached the seafloor. Calcium carbonate reacts readily with acids, as any field geologist knows well. A simple test for limestone, a rock composed principally of calcium carbonate, is to put a drop or two of weak acid on it and watch for a reaction. If it fizzes and dissolves, it is limestone. Rising levels of atmospheric carbon dioxide force more of the gas into the oceans, lowering pH, and seawater becomes more acidic. (This phenomenon has been detected in today's oceans, a result of the steady increase in atmospheric carbon dioxide over the past fifty years or so; seawater has become measurably more acidic, although as yet the change is much smaller than the shift that occurred during the PETM.)

The extent of calcium carbonate dissolution in PETM sediments is thus an indirect measure of the atmosphere's carbon dioxide content at that time, and, together with details of the carbon isotope shift, is the basis for calculations of the amount of carbon added to the PETM carbon cycle. Exactly how much of the several thousand billion tons indicated by this calculation was initially methane is still an open question, but all of it ended up as carbon dioxide, raising the atmospheric concentration to almost double the pre-PETM value (which was already high by today's standards). Seawater temperatures rose in step, and global average surface temperatures increased by at least nine degrees Fahrenheit, and perhaps as much as sixteen degrees.

Climate models incorporating all available data for the PETM make it clear that although the interval was initiated by the rapid release of a very large amount of carbon, it was sustained by continuing emissions, either sporadically or continually. They also suggest that there were feedback processes, not yet fully understood, that amplified the greenhouse warming. An intriguing suggestion has been made recently that could explain these and other perplexing aspects of the PETM record. It is based on two very different sets of observations: the behavior of lakes in volcanic regions, and the workings of plate tectonics in the North Atlantic Ocean around the time of the PETM.

In 1986 nearly two thousand people and twice as many livestock suddenly and mysteriously died near Lake Nyos, in Cameroon. There was no epidemic, no large earthquake, no volcanic eruption or severe storm. What soon became apparent was that the deaths had been caused by asphyxiation. A burst of carbon dioxide gas, heavier than air, had been released from Lake Nyos, and the deadly cloud had spread through low-lying terrain around the lake and had flowed down adjoining valleys, asphyxiating every living thing in its path. Like many deep lakes, Lake Nyos is stratified, with warm, low-density surface waters and cooler, heavier water at depth, a condition that prevents mixing. But Lake Nyos occupies an old volcanic explosion crater, and volcanic gases continually seep upward into its deep waters. In 1986, something—perhaps a landslide or a small earthquake—disturbed the lake's stratification, causing deep water, saturated with carbon dioxide, to rise. As it reached shallower, lower-pressure regions, the carbon dioxide came out of solution, expanded, and literally exploded out of the lake's surface in a fountain of gas and water at least three hundred feet high. The best estimates indicate that more than one and a half million tons of carbon dioxide were released in this single episode.

Could something similar—but on a much greater scale—be responsible for the PETM? In the North Atlantic of 55 million years ago, the strait separating Norway and Greenland was much narrower than it is at present (seafloor spreading has pushed the two continents farther apart since that time). When volcanic activity thrust the seafloor upward in the southern part of the strait, a deep, isolated basin was created, cut off from communication with the Atlantic Ocean. In the global warmth of the time, both Greenland and Norway were heavily vegetated and the precipitation level was high, so abundant organic debris washed into the basin. Bacteria in the sediments generated large amounts of methane, and carbon dioxide seeped into the deep water from the volcanic magmas below. Both gases would have built up to high levels in the bottom waters—if the basin was stratified.

The argument about stratification is still speculative—there is no obvious clue in the geological record that either confirms or disproves the possibility. But a case can be built on plausibility. At the time of the PETM, the basin was more than a kilometer deep, and because it was isolated from the Atlantic Ocean, water circulation was restricted or nonexistent. A large density difference between surface and deep water is crucial for stratification to occur, and in the prevailing warm climate—even at the latitude of Greenland, surface temperatures were around seventy degrees Fahrenheit or higher—surface water would have been warm and so would have had a low density. Fresh water runoff into shallow regions, and a supply of dense, salty water at depth from volcanic springs, would have further strengthened the density contrast. Under such conditions, methane and carbon dioxide concentrations would have built up toward saturation in the bottom waters of the basin, until, as at Lake Nyos, some event disrupted the stratification and initiated a turnover, with explosive release of the dissolved gases. The basin was about the size of the Red Sea today, and although it is not easy to estimate the exact volume of methane and carbon dioxide in the saturated bottom water (the estimate depends on details of the basin's topography, which are poorly known), straightforward calculations indicate that at least 100 billion tons of methane would have been present. When released, this would have been more than enough to initiate the documented PETM temperature rise. Stratification would have been restored following this initial explosion, and once again gases would have begun to accumulate in the deep waters of the basin. The cycle would have repeated itself, perhaps on a timescale as short as a few decades.

This scenario for kick-starting the PETM global warming was proposed by an international group of researchers from Britain, Ireland, and France in a 2009 article in the journal *Nature*. One of the attractions of the idea is that it does not posit the destruction of methane hydrates worldwide by a catastrophic event, which has always seemed unlikely. The authors of the *Nature* paper compare the isolated North Atlantic basin to a capacitor that repeatedly builds up an electric charge and then releases it. Injections of tens of billions of tons of methane from the stratified basin every few decades over several thousand years, accompanied by release of an unknown quantity of carbon dioxide, would explain the rapid and sustained temperature increase of the PETM. Continued operation of this mechanism, even at a diminished level, would also account for the persistence of high temperatures over more than 100,000 years. The "methane capacitor" would cease to operate—and global temperatures would begin to drop—when communication between the basin and the waters of the North Atlantic was reestablished.

There are other theories about the cause of the PETM, too—even some that suggest sources besides methane for the added carbon. One involves the impact of a comet containing large amounts of carbon of the "right" extreme isotope composition. However, there is no credible supporting evidence, such as a suitable impact crater, for this idea. Other scenarios invoke the injection of massive quantities of carbon dioxide into the atmosphere through rapid oxidation of large amounts of organic carbon stored in sedimentary deposits. For example, one idea is that peat deposits, which were widespread 55 million years ago, burned in a kind of global conflagration because of a drying climate. The extensive 1997 forest fires in Indonesia, brought on by drought related to an extended El Niño condition, served as a model for this hypothesis. The Indonesian fires generated the equivalent of about a year's worth of human-produced carbon dioxide in a relatively short time, not an inconsiderable amount. However, we don't really know if the PETM climate was truly dry enough to support global peat burning. Moreover, the "ordinary" organic carbon of peat has a much less extreme isotopic composition than biologically generated methane, so the amount of carbon from this source required to explain the isotope shift is unreasonably large.

In addition to the problem of the PETM carbon source, there is another aspect of this warm period that puzzles earth scientists. Evidence from the geological record indicates that the carbon dioxide content of the atmosphere was near 1,000 parts per million by volume just before the PETM began (compared to a present-day level of 385 parts per million). During the PETM, the concentration nearly doubled, rising to 1,700 or 1,800 parts per million. This has posed a problem for climate modelers because the current understanding of the "sensitivity" of climate to increases in carbon dioxide, a crucial parameter for predicting future climate, is that a doubling of the atmospheric concentration of carbon dioxide should lead to a temperature rise of three to, at the very most, eight degrees Fahrenheit. The consensus view is that the increase would be toward the low end of this range. Yet even though carbon dioxide increased by only a factor of 1.7 or 1.8 during the PETM, global surface temperatures rose by nine to sixteen degrees. Clearly, this has implications for forecasting the climate impact of future greenhouse gas increases. Are there feedback mechanisms we don't know about? Does it mean that the warmer world of 55 million years ago, with no polar ice, is a poor analog for understanding present-day conditions? Unfortunately, there are still no clear answers to these questions.

However, we do know quite a bit about the effects of the PETM climate changes on flora and fauna, and these may be a good guide to the biological consequences of current global warming. As pointed out at the beginning of this chapter, a characteristic of the PETM was a high rate of extinction among some planktonic organisms, specifically the foraminifera. But the extinctions were selective, predominantly affecting deep-dwelling foraminifera; between 35 and 50 percent of these species disappeared, the largest extinction to strike this particular group in the past ninety million years. Foraminifera living in near-surface water were least affected; their geographical distribution patterns changed in response to higher seawater temperatures, but extinction rates did not increase significantly. There are several possible reasons for the differing effects on the deep- and shallow-dwelling organisms. For example, calcium carbonate dissolves more readily in cool, deep water, so it is possible that seawater became so acidic at depth that the deep-dwelling foraminifera could not maintain their shells. Also, deep-water foraminifera, accustomed to low water temperatures, may have been slower to adapt to the warm temperatures than their surface-dwelling cousins. Or, as some have suggested, oxidation of large amounts of methane injected into deep waters may have so depleted the deep ocean of oxygen that the foraminifera were killed off. All of these scenarios are plausible, but none has yet been proven.

On land, the fossil record is less complete, but it too reveals major changes in flora and fauna during the PETM. Long before this warm interval was discovered, it was known that there had been significant "turnover" among Northern Hemisphere mammals at the beginning of the Eocene epoch. Careful comparisons between fossil and isotope records show that this change coincided precisely with the carbon isotope excursion that defines the PETM. In the western United States, for example, many new "immigrant" mammal species suddenly appeared as the PETM began. The immigrants quickly came to dominate the ecology of the region, and the overall diversity of mammal species increased. Unlike the elevated temperatures of the PETM, the biological changes were permanent; when the climate returned approximately to its pre-PETM conditions, the new mammalian ecosystem remained. Plants too reacted to the climate changes, mostly by migration to higher latitudes and altitudes in response to increased temperatures. This pattern resembles the changes that have occurred during the glacial-interglacial episodes of the Pleistocene Ice Age, when Northern Hemisphere plant species have repeatedly marched north during the interglacials and south during the glacial intervals. Biologists are beginning to see the same pattern in response to the comparatively minor global warming of the past half-century or so: this small temperature rise has resulted in documented shifts to higher elevations and latitudes for some species.

Perhaps the most important message to emerge from the fossil record of the PETM, both marine and land-based, is that even such a geologically transient climate event, lasting little more than 100,000 years, can have major biological consequences, and that those changes can persist long after the event itself comes to an end. Some of the effects of the PETM—for example, those among the mammals—are still apparent 55 million years later.

When the various records of change through the PETM are examined closely, it becomes obvious that the interval had several distinct phases. This is particularly clear from the carbon isotope values (see figure 26). There was an initial rapid decrease in these values, coupled with a sharp increase in global temperatures. This was followed by a period of relative stability lasting about sixty thousand years, when carbon isotopes remained roughly constant, at much lower values than before the PETM, and then, during the recovery phase, isotope values rose quickly and temperatures fell. Other parameters, such as the calcium carbonate content of sediments, follow roughly the same pattern.

Understanding the reasons for these distinct phases is important, because they hold clues to the operation of the Earth's climate system. There is little dispute about the initial period of rapid change; as we have already seen, it was the result of the injection of a large amount of carbon into the ocean-atmosphere system. The cause of the relatively long interval of stability, however, is less certain. Many geoscientists think that the initial carbon addition and consequent rapid warming must have jolted the Earth out of one stable climate state and into another. The idea that various forcing factors can nudge the climate from one stable mode to another, sometimes expressed in terms of a tipping point, has gained credibility from study of more recent abrupt climate changes, such as the Younger Dryas period discussed in the previous chapter. During the approximately sixty-thousand-year period within the PETM when carbon isotope values were stable, however, there were other changes taking place, albeit gradually. For example, increasing amounts of calcium carbonate appear in some sediment cores, indicating that the acidity of the oceans had started to decrease.

During the recovery phase at the end of the PETM, all of the key parameters began to change much more rapidly. Carbon isotope values rose, temperatures dropped, and calcium carbonate deposition returned to its pre-PETM condition. What processes facilitated this recovery? Dissolution of calcium carbonate slowly started to neutralize the ocean's acidity during the stable phase of the PETM. Increased weathering on land began to reduce the high carbon dioxide concentration of the atmosphere, decreasing the greenhouse effect and lowering temperatures. The various feedback mechanisms of the Earth's carbon cycle kicked in to force the system back toward its original state after the shock of massive carbon injection. More evidence will have to be pried out of the geological record before the workings of these processes can be fully quantified, but the general outline is reasonably well understood.

What lessons about possible future climate change can we take from our current knowledge of the PETM? One is that present-day carbon emissions, which are mainly from fossil fuel burning, are truly unprecedented. Throughout this chapter, I have emphasized that the PETM was initiated by a massive and rapid injection of carbon into the ocean-atmosphere system. But "massive" and "rapid" are relative terms. The carbon isotope value in ocean sediments decreased over a period of twenty to thirty thousand years at the beginning of the PETM, which is probably a rough indication of the timescale of carbon injection (to be fair, all indications are that carbon addition was far more rapid at the beginning than at the end of this initial period). Over the next two centuries, however, if carbon emissions continue at approximately their current rate, something like 4,000 to 5,000 billion tons of carbon—about the same amount that was added during the *entire* PETM—will be injected into the Earth's surface environment by humans. The timescale is hundreds of years, not thousands or tens of thousands, making the rate of carbon addition at least ten times more rapid than during the PETM.

Because the rise of atmospheric carbon dioxide from human activity is so fast, and because its cessation will also be abrupt when we run out of fossil fuels (or, more hopefully, when non-fossil-fuel energy sources take over), the duration of very high temperatures will be shorter than during the PETM. Climate models indicate that most of the man-made carbon dioxide will be absorbed by the ocean, and that atmospheric concentration, as well as temperature, will decline rapidly (geologically speaking) once the peak of emissions has passed. However, because of the intricacies of the carbon cycle, carbon dioxide levels will likely stabilize at a value higher than today's for an extended period of time—many tens of thousands of years—and temperatures will also remain higher than they are now.

These conclusions are based solely on consideration of carbon dioxide and the carbon cycle. They do not include potential amplifying effects, such as the possibility that rapidly increasing temperatures will destabilize methane hydrates currently locked in Arctic permafrost or buried in continental shelf sediments. Release of methane from these sources, particularly if it is rapid, could intensify and prolong the man-made warmth.

Current climate models predict that polar regions will heat up faster than low latitudes, primarily due to the fact that they will absorb more heat as albedo decreases because of melting ice. Temperature records from the past few decades indicate that this is already happening. At the time of the PETM, the polar regions were essentially ice-free and would have experienced only minor changes in albedo. Even so, evidence from high-latitude sediment cores shows that temperatures in these regions shot up and remained high—higher than predicted by climate models, unless unreasonably high concentrations of greenhouse gases are invoked. This suggests that there were feedback processes in operation that are still not understood, and it raises the possibility that the climate of high-latitude regions will change even more drastically than currently anticipated.

Evidence from coastal sediment cores, as well as from freshwater sediments in continental interiors, documents significant changes in precipitation during the PETM. Especially at middle and high latitudes, there are abundant indications of much higher rainfall rates, and, from some European regions, evidence of frequent large-scale flooding. This is consistent with the increased evaporation and generally more humid climate expected under warm conditions, and the resulting speedup of the entire hydrological cycle. In contrast, the climate became drier in the western United States for part of the PETM. This meshes with the results of most climate models that simulate the effects of man-made global warming, which predict large but very complex changes in the hydrological cycle, making it difficult to foretell regional effects in detail.

The Great Warming of the PETM parallels, in some respects, shifts in the Earth's systems that are occurring today; it was a natural experiment that provides valuable insights into what the long-term effects of human-induced changes may be. Much has already been learned about its probable causes and global consequences, as should be obvious from the discussion in this chapter. But paleoclimatologists continue to search for further clues about this event in the geological record, and if they are clever enough to decode them, that will help make forecasting the Earth's future climate more accurate—the ultimate goal of those involved in this type of research.

CHAPTER TEN

Reading LIPs

Scientists are fond of acronyms, the catchier the better; and geologists are no exception. The "PETM" of the previous chapter isn't an especially inspired example, but in 1993 a group of geologists interested in a particular kind of volcanism came up with an acronym they thought would nicely describe the objects of their fascination: LIPs. They formed an international organization to study them, and they even feature a "LIP of the month" on their Web site. But these LIPs are not what someone stumbling across the Web site might think. In the geological context, the acronym stands for *large igneous provinces*.

Geological LIPs are produced during periods of intense volcanic activity when huge volumes of rapidly flowing lava are erupted onto the surface of the Earth and spread out over vast areas. LIPs are often referred to as "flood basalts" because the lava flows that produce them literally flood the existing landscape. Individual basalt flows in some of these provinces can be traced for two hundred miles or more, flow after flow can pile up to a thickness of several miles, and the largest of the provinces extend over millions of square miles. The volcanism that produces LIPs usually occurs over a geologically short period of time; for those provinces that have been studied in detail, most of the lava was erupted within a few hundred thousand to a million years.

The consensus is that LIPs originate—like Hawaii and its trailing island chain—when massive plumes of hot mantle material rise from deep in the interior, melt, and burst through the overlying crust. Illustrations of these plumes often show them looking like vertical tadpoles, with a large head butting up into the lithosphere and a long, narrow tail extending deep into the mantle. LIPs mark the initial voluminous eruptions from the head of the plume, but the tail can continue to feed volcanoes for many tens of millions of years, although at a much reduced rate. One well-known LIP is the Deccan Plateau, a miles-thick stack of basalt flows in central India. It was formed over a short period around 65 million years ago as the Indian continent moved northward over the head of a plume. Today the volcanically active island of Réunion in the southern Indian Ocean is believed to tap the thin tail of the same plume.

LIPs occur both on the continents and on the seafloor. Unlike most other kinds of volcanism, they have no particular relationship to plate boundaries, because the plumes that form them originate at great depths, far below the base of the plates. The most extensively studied LIPs are those on land; they include the Deccan Plateau, the Siberian flood basalts in Russia, the smaller Columbia River Plateau in the Pacific Northwest of the United States, and a number of others. The Siberian basalt province, which erupted just over 250 million years ago, is the oldest of the well-preserved LIPs. The surface flows in most examples that are older than this have been severely eroded or even completely weathered away; however, their one-time presence can be inferred from the existence of numerous volcanic conduits through the crust that were, at the time of eruption, deeply buried but have since been exposed at the surface through uplift and erosion.

Although the acronym is a recent invention, LIPs have been known on the continents ever since scientists began examining rock formations in detail. It is only with increased exploration of the seafloor, however, that numerous undersea LIPs have been discovered. The largest of these is the Ontong-Java Plateau in the western Pacific Ocean, formed by a massive outpouring of lava about 125 million years ago that built up the ocean crust in that region to many times its normal thickness. Another ocean-floor LIP is in the Caribbean. There the major pulse of volcanism occurred during the second half of the Cretaceous period, about 93 to 94 million years ago. The plume that created the Caribbean LIP may still be active; plausible reconstructions of how the tectonic plates in the region have moved over the past 100 million years suggest that the volcanoes of the Galápagos Islands now sit above the site of this plume.

If you could go back in time and visit the Earth when the Caribbean LIP was erupting, you would find a warm climate engulfing the planet from pole to pole. The greenhouse effect was strong; the carbon dioxide content of the atmosphere was much higher than it is today. Palm trees grew north of the Arctic Circle, and there were no glaciers on Greenland and no massive ice cap in the Antarctic. The outline shapes of the continents would be as unfamiliar to you as the climate, because—with little water tied up in glaciers—sea level was high; large portions of Europe and southwestern Russia were under water, and a broad, shallow sea ran from the Gulf of Mexico to the Arctic, splitting North America in two (see figure 27). The locations of the continents would seem strange, too. Australia and Antarctica were still joined together in a single continent, and India, on its way north toward Asia, was an island off the east coast of Africa, sitting next to Madagascar. The Atlantic Ocean was still in the early stages of formation; only a narrow sea separated Brazil from the west coast of Africa. Dinosaurs dominated animal life on land.

The high carbon dioxide content of the Cretaceous atmosphere can be ascribed to a disparity in nature's balancing act, the carbon cycle: volcanic input of carbon dioxide outweighed its take-up by weathering and other processes. The Atlantic Ocean was opening up, the large southern continent of Gondwana was fragmenting, and seafloor spreading was rapid, with large volumes of basalt erupting on the seafloor and bringing with it carbon dioxide from the Earth's mantle.

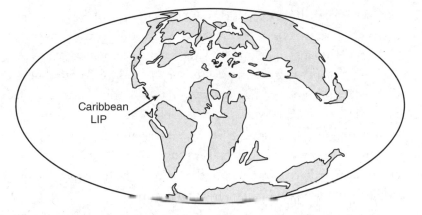

Figure 27. The continents as they appeared at the time of the Caribbean LIP eruptions, 93 to 94 million years ago. Sea level was high; much of North America and Europe was under water, and there were shallow seas over parts of Africa and South America. (After a map by Ron Blakey, Northern Arizona University; see http://jan.ucc.nau.edu/~rcb7/.)

Some of the excess atmospheric carbon ended up stored in ocean sediments; sedimentary rocks from the Cretaceous are notably rich in organic carbon (they are the source of much of our oil and gas for this reason). In some places there is so much carbon that the rocks are quite literally black. These especially carbon-rich layers, known as "black shales," form only when there is abundant plant and animal life in the surface waters, providing a constant rain of dead organisms to the seafloor, but they also require that the deep water contains little or no dissolved oxygen; without oxygen, the organic matter can be preserved in the sediments, unoxidized. Today such conditions exist only in localized environments, such as deep fjords where the bottom water is stagnant and oxygen-starved, or the Black Sea, which has nutrient-rich and highly biologically productive surface waters but stagnant and oxygen-poor deep waters.

Though Cretaceous black shales had previously been documented in many places on land, it was only in the 1970s, through the work of the Deep Sea Drilling Project (DSDP), that the global extent of these deposits was recognized. Throughout the world's oceans, wherever DSDP cores brought up samples of Cretaceous age, there were layers of especially organic-rich sediments. Some of these layers were thin and localized, but three were found in nearly every sediment core drilled and could be correlated with prominent black shale deposits on land. Subsequent work has shown that these three intervals record times of major, worldwide environmental change. Because their deposition required low-oxygen ocean waters, they have come to be known (using another acronym) as OAEs—oceanic anoxic events.

Evidence from sedimentary rocks indicates that both shallow and deep waters of the ocean have been oxygenated almost continuously since late in the Proterozoic eon, when the oxygen content of the atmosphere rose sharply to near present-day levels. The three Cretaceous OAEs, however, signify short periods when this was not the case. During these intervals, deep water throughout the oceans became starved of oxygen. Furthermore, the transitions from normal sediments to the organic-rich black shale layers and back again for these OAEs are abrupt. What plausible mechanisms could create oxygen-poor deep-ocean water worldwide, and also be capable of causing rapid switches between oxygenated and anoxic conditions?

One of the first hypotheses put forward was that the Cretaceous OAEs were the result of a strongly stratified ocean with stagnant bottom water, its oxygen consumed in reactions with sinking organic matter and not replaced. In the present-day open ocean, circulation is driven by cold, dense, oxygen-rich surface water that sinks at high latitudes and continually refreshes the deeper parts. But in the very much warmer Cretaceous climate, even polar regions were mild. There was no cold, dense water at high latitudes to drive ocean currents, and circulation was, at best, sluggish. Dissolved oxygen at depth would have been quickly used up as sinking organic matter was oxidized. The situation may have been exacerbated by plate tectonics: the narrow Atlantic Ocean of the Cretaceous further restricted circulation.

The difficulty with this scenario is that it doesn't really explain either the sudden onset or the rapid end of the OAEs. Sluggish circulation and low oxygen in the deep ocean can explain the generally elevated amounts of organic material in Cretaceous sedimentary rocks, but something else is necessary to account for the OAEs.

One possibility is that volcanism and the release of carbon dioxide into the atmosphere, the same process that created the warm Cretaceous climate in the first place, could be responsible. But it would have to be a voluminous, short-lived burst of volcanism superimposed on the already high average rate of eruptions. Formation of a LIP would fit that criterion.

The youngest of the Cretaceous OAEs—the most intense of the three intervals, judging by the chemical properties of the black shales—occurred between 93 and 94 million years ago. It thus coincides in time with the formation of the LIP on the Caribbean seafloor, and also with one of the boundaries geologists use to subdivide the Cretaceous period, the Cenomanian-Turonian (C-T) boundary, named after the time intervals on either side (these names refer to geographical localities in France where rocks of this age were first described in detail). The boundary is marked by significant extinctions of marine organisms, predominantly bottom dwellers and those that lived in deep water, suggesting that the low oxygen content of the deep ocean was a factor in their demise.

Most of the volcanic rocks of the Caribbean LIP still lie on the seafloor, and the province as a whole remains poorly sampled. However, in a few places, including Haiti and the Lesser Antilles, basalt flows from the episode have been lifted up onto land by tectonic processes. These flows provide the most accurate and reliable ages for the initial, high-eruption-volume period of the province; they restrict it to the interval between 93 and 94 million years ago. A few younger dates have also been reported, including some from submarine rocks, and they range down to about 87 million years, suggesting that volcanism persisted for as much as six or seven million years. But the coincidence of peak volcanism in the Caribbean LIP with the OAE is striking and marks the LIP as a possible trigger for the environmental changes that precipitated black shale deposition and the concurrent extinctions.

There is additional circumstantial evidence linking the Caribbean LIP to the OAE. Several indicators in sedimentary rocks show that the already high surface-water temperature in the tropical Atlantic rose rapidly by at least an additional five degrees Fahrenheit and possibly by as much as fifteen degrees near the beginning of the OAE. Carbon dioxide released during the initial eruptions of the Caribbean LIP was most likely the cause. Also, the black shales from near the C-T boundary have high metal concentrations that may have originated from the erupting lavas.

The convergence of these various bits of evidence convinced many researchers that there is indeed a link between Caribbean LIP volcanism and the anoxic event at the C-T boundary. But the clues were all circumstantial; there was nothing really definitive to cement the connection. Then, in 2008, the proverbial smoking gun was uncovered. Steven Turgeon and Robert Creaser, both at the University of Alberta, found incontrovertible chemical evidence that massive undersea volcanic eruptions drastically affected ocean-water chemistry right at the C-T boundary. The only known large volcanic episode during that time period was the Caribbean LIP.

The discovery, like many in science, was partly serendipitous. Steven Turgeon is a geochemist, and one of his interests is to develop better ways to date sedimentary rocks. Turgeon doesn't use any of the better-known dating methods described earlier in this book, but instead employs an exotic (exotic to non-geochemists, at any rate) age-determination scheme based on the radioactive decay of the rare metal rhenium to an isotope of another rare metal, osmium. The organic-rich rocks near the C-T boundary were an appealing target because they were known to have high metal contents, and also because, if the method worked, it would provide valuable information about the absolute age and possibly the duration of the OAE. So Turgeon went off to Italy to collect material from one of the classic localities for C-T boundary rocks, and he also got samples of organic-rich sediments spanning a similar age range from a sediment core taken off the northeast coast of South America. The two locations are more than 3,500 miles apart today, and they were in different oceans at the time of the C-T boundary (see figure 27).

But the research didn't work out as planned. The method Turgeon was using is very labor-intensive, and when he finally assembled all his data after months of toiling in the laboratory, he found that the results were much too erratic to give him precise ages for the sedimentary rocks. It looked as though the whole project might have to be abandoned. However, Turgeon noticed that samples from the organic-rich black shale intervals in both the Italian rocks and the sediment core had much higher concentrations of osmium than he had expected. He thought this observation was interesting, but he wasn't sure what it meant. So he put the data aside and turned to other work, although he didn't forget about the black shales entirely. He kept wondering about the high osmium contents, and if there was another way to make sense of the measurements.

About half a year later, hoping to salvage something from the huge effort he had invested, Turgeon pulled out the data again. Perhaps there was something he could use for a short presentation at a scientific meeting. He tried excluding data points that seemed to be anomalous, but that didn't make much difference. In frustration, he decided that he should reanalyze all his samples just in case he had made a mistake the first time around. But the new results were the same as the earlier ones: there was no problem with his analyses; it was just that Mother Nature was putting roadblocks in the way of understanding the data. Then Turgeon had an idea. Instead of using his results to date the samples, he could turn the process around. The age of the C-T boundary (93.5 million years) was already well known, so he could calculate how much osmium had been produced in his samples by radioactive decay over the past 93.5 million years and subtract that number from the amount he had measured; the difference would allow him to calculate the original isotope ratio in his samples when they were first deposited. Because nearly all the osmium in sediments comes from seawater, he reasoned that the calculated ratio would be very close to the osmium isotope ratio of seawater at the time of the OAE.

When Turgeon made this calculation, he immediately realized he was onto something significant. He had measured samples from before, during, and after the black shale interval, and his data showed that the osmium isotopes in Cretaceous seawater had changed abruptly right at the beginning of the OAE, then returned to normal at the end. The Italian rocks and the ocean core samples from 3,500 miles away told exactly the same story: in terms of osmium, at least, there had been a rapid, worldwide change in seawater chemistry that coincided with the OAE.

In a sense, isotopes are like fingerprints. Geochemists use them in much the same way that detectives at a crime scene use fingerprints: to identify the culprit. The arguments can be complicated, but in many cases the ratio between two specific isotopes can be used to identify an event or a process uniquely. We saw in the previous chapter how carbon isotope ratios were used in this way to identify biologically produced methane as the source of carbon during the Great Warming of the PETM. Some isotopes are better for these sorts of tasks than others, and osmium is especially good, because there are very large differences in osmium isotope ratios between rocks formed in different ways—for example, between rocks from the continental crust and those formed by ocean-floor volcanism. Both of these sources contribute osmium to the oceans, and the resulting isotope ratio of the osmium dissolved in seawater depends on how much has come from the continental crust by weathering, and how much was supplied by ocean floor volcanic activity. The sudden shift that Turgeon found indicated that there had been a huge increase in the amount of volcanic osmium in seawater at the beginning of the OAE.

According to Turgeon's data, the volcanic osmium made up about 97 percent of all osmium dissolved in the ocean throughout the black shale interval, between thirty and fifty times more than before the event. Today less than a third of seawater osmium originates from undersea volcanic activity; the bulk of it comes from erosion of the continents. Turgeon couldn't *prove* that the volcanic osmium came from the Caribbean LIP, but there is no other obvious candidate that can explain the data.

Turgeon and Creaser's work thus makes a direct link between the intense volcanic activity of the Caribbean LIP and a period of global ocean anoxia, widespread marine extinctions, and the accumulation of organic-rich sediments on the seafloor. It provides a solid foundation for the idea, until recently based primarily on circumstantial evidence, that the periodic volcanic outpourings of LIPs have played a significant role in the Earth's history, including triggering global climate change and extinctions. From all available evidence, it appears that the most important aspect of LIPs in this regard is the large amounts of carbon dioxide released during the eruptions. That is an important conclusion, because it means that sedimentary rock strata laid down during these periods of large-scale volcanism, just like the rocks from the PETM interval described in the previous chapter, may preserve clues about how the planet responded in the past to rapid changes in atmospheric greenhouse gases.

The period of ocean anoxia linked to the Caribbean LIP was one of the most intense in the geological record of the past few hundred million years, and for that reason it has received a lot of attention. Geochemists have managed to unravel many details of what happened during the interval by measuring "proxies" for various environmental characteristics in sedimentary rocks from that time (oxygen isotopes in plankton shells, for example, are a proxy for water temperature). These show that at the beginning of the OAE, as we saw earlier, water temperature warmed by several degrees almost immediately in response to the initial burst of volcanic carbon dioxide. Interestingly, it didn't stay high for long: not far into the interval, seawater cooled again, with temperatures dropping back to levels near those at the beginning. Almost certainly, this was due to extraction of carbon dioxide from the atmosphere through the carbon cycle: photosynthesizing organisms proliferated in the warm surface waters of the ocean, using carbon dioxide to make organic material that, when the organisms died, was buried in the sediments and stored there. The effectiveness of greenhouse warming decreased, and temperatures declined. But even the enormous sequestration of carbon caused only a short (less than about 150,000 years) dip; the continued input of carbon dioxide from Caribbean LIP volcanism soon tipped the balance, and temperatures rose again.

One proxy, a biomarker specific to certain types of bacteria, has recently shed further light on ocean conditions during the OAEs. Although the host organisms are not preserved as fossils, hardy molecules unique to "green sulfur bacteria" are common in the black shales from the C-T boundary. These bacteria are tiny, interesting creatures that don't require oxygen to live; instead, they obtain energy by consuming the toxic gas hydrogen sulfide. But they are also photosynthetic and live in shallow water where they are exposed to sunlight. Their importance, in terms of working out conditions at the C-T boundary, is that they exist only in a very specific environment, at the interface between deep water rich in hydrogen sulfide and sunlit, shallow water that contains dissolved oxygen. Whenever biomarkers for green sulfur bacteria are found in sedimentary rocks, they characterize a past ocean in which significant amounts of hydrogen sulfide pervaded deep regions and percolated upward to invade the near-surface waters.

Most green sulfur bacteria today inhabit stagnant, oxygen-poor bodies of water like the Black Sea. They live on hydrogen sulfide produced by another variety of bacteria dwelling in the deeper, oxygen-free water that characterizes such environments. These deep-dwelling bacteria break down sulfate molecules (an abundant component of seawater and Black Sea water) to obtain energy, releasing hydrogen sulfide in the process. The green sulfur bacteria, in their near-surface niche, consume it as it permeates upward. The OAE sediments containing biomarkers for green sulfur bacteria were deposited in deep, open-ocean water, signifying that much of the world's deep ocean was inhabited by hydrogen-sulfide-generating bacteria at that time. The deep sea was suffused with the toxic gas, and some of the marine extinctions at the C-T boundary may have been caused by its deadly effects.

In the roster of known LIPs, the Caribbean eruptions were relatively small. I've dwelt on this episode at length here because the osmium isotope signature in the OAE black shale ties the volcanism very firmly to global environmental effects. What, then, were the consequences of much larger LIP eruptions? In most cases, the arguments have to be made on the basis of coincidence in timing. But as radiometric dating of the volcanism on the one hand and environmental changes on the other has gotten more precise, links between the two are becoming clearer. In particular, three of the largest mass biological extinctions of the past 500 million years coincide closely with the formation of three major LIPs: the Deccan flood basalts of central and western India, the Central Atlantic Magmatic Province (CAMP) LIP, and the Siberian flood basalts. All three provinces are anomalous in terms of the sheer volume of lava erupted over a relatively short period of time. And each of them coincides with a major geological boundary: the Deccan basalts with the end of the Cretaceous period, the CAMP LIP with the end of the Triassic, and the Siberian basalts with the end of the Permian. The early geologists who set these boundaries did so entirely on the basis of sharp changes in the fossil record; they had no inkling of the coincidence in timing with massive volcanic outpourings.

If the timing of only one LIP overlapped with a major extinction episode, it might be dismissed as coincidence. But with three examples in the past 250 million years, a connection is highly likely, even if it is indirect. What could the linkages be? This is an area of much current research, but many earth scientists consider volcanism-related greenhouse gases, and consequent temperature changes, to be the most likely ultimate culprit—as seems to be the case for the less severe Caribbean LIP. Based on the total volume of lava, the eruptions of each of the three large LIPs released massive amounts of carbon dioxide, but a crucial question concerns how rapidly this occurred. If the eruptions spanned several million years, the carbon cycle would have kept atmospheric carbon dioxide roughly in balance, with relatively little temperature change. If the timescale was much shorter, rapid and substantial temperature changes would have followed. Compounding the direct and indirect effects of high temperatures, high atmospheric carbon dioxide would have caused ocean acidification, making life difficult for many marine organisms.

Evidence from the geological record shows that not all of the greenhouse gases entering the atmosphere during the periods of LIP formation came directly from the volcanic lavas. There are clues, for example, that the CAMP episode destabilized methane hydrates, adding methane to the atmosphere. There is also evidence that interactions between the hot magmas of the Siberian flood basalts and the sedimentary rocks they intruded forced the release of greenhouse gases. Regardless of the exact mechanisms, however, carbon isotopes in sedimentary rocks show that substantial amounts of carbon dioxide were added to the ocean-atmosphere system coincident with the eruption of each of these LIPs.

The Deccan flood basalts were erupted 65 to 66 million years ago, at the time of the K-T (end of Cretaceous) boundary. Flow upon flow of basalt flooded onto the Indian continent; these rocks, still more than a mile thick in places in spite of tens of millions of years of erosion, today form an impressive plateau in central India. But their possible relationship to the K-T extinctions has been overshadowed by the impact theory described in chapter 3. There is little doubt that the environmental aftermath of the K-T collision played a large role in these extinctions. But it may be that they would not have been as severe in the absence of the Deccan LIP eruptions. Dating studies show that the eruptions began somewhat before the boundary (and the collision), and it is likely that their environmental effects had already begun to stress life on Earth before the knockout punch of the impact. The volcanism also continued after the collision, perhaps further exacerbating its effects.

Geologists have probed rocks from all of the major mass extinction boundaries for evidence of impacts, but the K-T remains the only proven example. This strengthens the case that environmental change related to LIPs may be responsible for extinctions at other boundaries. Extinctions at the end of the Triassic period, 200 million years ago, were less extensive than those at the K-T boundary, but still it is estimated that more than 50 percent of all existing animal and plant genera disappeared. The extinctions were essentially simultaneous with formation of the CAMP LIP, which recent research shows took place over a period of about 600,000 years. The CAMP lavas, which were erupted as the supercontinent Pangea broke up and the Atlantic Ocean began to form, are estimated to have once covered an area approximately equivalent to that of the United States, and the associated carbon dioxide emissions were huge. CAMP lavas can be found today all along the margins of the Americas, Europe, and Africa.

However, the largest extinction of all took place 251 million years ago, at the boundary between the Permian and Triassic periods. The best estimate is that it wiped out about 95 percent of all life in the oceans and at least 70 percent of land dwellers. Nothing was sacred; reptiles, amphibians, plants, insects, fish, plankton, and shellfish all suffered severe extinctions. Life on Earth took a long time to recover, and the pattern of extinctions and survivals reverberates today, a quarter of a billion years later. This has been dubbed "the mother of all extinctions"—and it coincides with the eruption of one of the largest LIPs known from the geological record, the Siberian flood basalt province.

Radiometric dating shows that the massive volcanic outpourings of the Siberian LIP peaked over a period of about a million years, beginning shortly before the main Permian-Triassic extinctions. The Siberian LIP is unusual, because it includes explosive volcanism in addition to the fluid basalt lava that normally characterizes such provinces. Some researchers have suggested that dust and aerosols thrown up by violent eruptions could have blocked out sunlight, lowered temperatures globally, and played a role in the extinctions. This possibility has been reinforced by the observation that single explosive eruptions in the recent past have produced large quantities of high-altitude aerosol particles, leading to small but measurable global temperature decreases. Still, most geoscientists doubt that the cooling, even from sustained explosive volcanism, would have been severe enough and sufficiently long-lived to be responsible for the Permian-Triassic extinctions. Also, the large amounts of carbon dioxide released during the eruptions would have had the counteracting and longer-lasting effect of *raising* temperatures.

The search for details of how the Siberian eruptions might have led to such massive extinctions led Lee Kump and Michael Arthur, from Pennsylvania State University, and their colleague Alex Pavlov, from the University of Colorado, to an interesting possibility. They began with the assumption that the Siberian volcanism released vast amounts of carbon dioxide rapidly, which raised global temperatures. As the oceans warmed, they could hold less and less oxygen (like most gases, oxygen is more easily soluble in cold water than in warm), and the normal oxidation of sinking organic matter kept the oxygen content of the deep sea very low. Kump and his colleagues noted that some sedimentary rocks from near the Permian-Triassic boundary contain the same biomarkers for green sulfur bacteria that are found in the organic-rich sediments of the Cretaceous OAEs, indicating that at least parts of the ocean were permeated by hydrogen sulfide. No seafloor older than about 200 million years still remains, so it is not possible to drill into ocean sediments from the Permian-Triassic boundary to determine whether there were intervals of global anoxia like those of the Cretaceous. But some sedimentary rocks now exposed on land with ages close to 251 million years (the time of the Permian-Triassic boundary) contain black shale layers. With firm evidence for the presence of hydrogen sulfide, and indications that at least some parts of the ocean experienced periods of anoxia, Kump and his colleagues began to wonder whether hydrogen sulfide might have been produced in such large quantities in the deep ocean that it permeated right to the surface and leaked out into the atmosphere, with lethal effect.

Their calculations showed that such a scenario is well within the realm of possibility; it would not require an extraordinary increase in the population of deep-ocean hydrogen-sulfide-producing bacteria to tip the balance. Especially in areas where there was already modest upwelling, water saturated with hydrogen sulfide would rise to the surface, overwhelm any oxygen dissolved in surface water, and some of the gas would escape into the air. Kump and his colleagues calculated that even if this happened over only about one-tenth of 1 percent of the ocean surface, the quantity of hydrogen sulfide transferred into the atmosphere would be enormous, several thousand times the amount released from volcanic activity today. At such levels, it would have been highly toxic to many, if not all, land-dwelling organisms, and it could account for the Permian-Triassic extinctions.

There would have been other consequences, too. The presence of so much hydrogen sulfide would have inhibited the mechanism that normally breaks down methane in the atmosphere, causing its concentration to increase rapidly and to stay high. This would have further raised already high temperatures through the greenhouse effect, and because the toxicity of hydrogen sulfide increases as the temperature goes up, its efficacy as an agent of extinction would also have increased. To top off this already quite horrible-sounding list of possibilities, Kump and his colleagues added something else: they noted that high levels of hydrogen sulfide in the atmosphere would have destroyed the Earth's protective ozone layer. Without that shield, the Earth's surface would have been bathed in high levels of dangerous ultraviolet radiation.

But did these science-fiction-sounding scenarios—oceans without life-sustaining oxygen, toxic gas belching out of seawater and poisoning plants and animals, ultraviolet radiation blasting the surface—really happen? The rock record, as ever, should provide clues. The biomarkers for green sulfur bacteria confirm the presence of hydrogen sulfide in the oceans. Other evidence is less direct, but seems to support the predictions made by Kump and his colleagues—if not in every detail, then at least in a general sense.

One clue comes from fossil plankton, which are numerous, evolve quickly, and can be traced from rock layer to rock layer with high time resolution. At the Permian-Triassic boundary, extinctions among the plankton appear to have occurred in pulses spread over several hundred thousand years. This is in distinct contrast to the extinctions at the Cretaceous-Tertiary boundary, which are—in a geological sense—essentially instantaneous. At the Permian-Triassic boundary, fossils of land vertebrates collected from lake and stream sediments, although much rarer than the plankton fossils, also show a more prolonged period of extinctions. This pattern is consistent with a slowly deteriorating environment hit by periodic triggering events—such as hydrogen sulfide gas filling the ocean and escaping into the atmosphere—that occurred in pulses. LIP volcanism, spread out over a million years or so but occurring in short, massive eruptive bursts separated by thousands or even tens of thousands of years of less intense activity, is a likely ultimate cause.

However, perhaps the most telling evidence comes from fossil spores of land plants. Before the Permian-Triassic boundary, woody, forest-forming plants were dominant, but they suffered heavy losses during the extinctions. Smaller, more primitive plants related to present-day mosses moved in to take their place. The spores produced by these usurpers are very resistant to degradation, and large numbers of them have been preserved as fossils. The most interesting thing about these fossil spores is that many of them exhibit the kinds of mutations that are produced by exposure to ultraviolet radiation. This suggests that the ozone layer was weakened or absent when these plants grew. Malformed spores are found worldwide in rocks from near the P-T boundary, and a particularly detailed study of fine-grained sedimentary rocks from Greenland shows that their abundance varies over time, exhibiting two distinct peaks separated by a few hundred thousand years. This pattern is consistent with an extended period of Siberian LIP volcanism punctuated by short, intense episodes that triggered hydrogen sulfide release from the ocean and destruction of the ozone layer.

The accumulated evidence from the geological record indicates that something like the bizarre scenario proposed by Kump and his colleagues may really have happened, and also confirms the general conclusion that LIP events can have global environmental effects even though they are confined to relatively small areas of the Earth's surface. The evidence of LIP-related extinctions discussed in this chapter also emphasizes a crucial common thread running through many aspects of the Earth's history: the importance of greenhouse gases, particularly carbon dioxide, as a primary agent driving widespread environmental change. In many instances, increasing levels of carbon dioxide in the atmosphere apparently reached a threshold at which the attendant global warming triggered other, follow-on processes that were the actual causes of the large-scale biological extinctions. Although we have only a rough idea of what those thresholds were, the best current estimates put the carbon dioxide content when these crises began somewhere between about two and a half and five times present levels. In one sense, that might seem to be a comforting conclusion, because human activity has increased carbon dioxide levels by only about a third since the beginning of the Industrial Revolution. Unless another LIP comes along—which is unlikely in the short term—it will take a long time to get to really dangerous levels.

But will it really? The rate at which greenhouse gases are being released into the atmosphere is increasing, and without a major effort to reduce emissions, the low-end levels at which past OAEs have occurred could be reached in just a few centuries. That won't affect you or your children, but two or three hundred years is not really such a long time in human history. Some people born today will live through a third to a half of that time span. We can only hope that the stark lessons from the geological past will push us all into action so that our descendents, a few centuries hence, will not be blaming us for setting the planet on a course toward ecological catastrophe.

CHAPTER ELEVEN

Restless Giants

Until recently, the Aeta people on the Philippine island of Luzon, descendents of seminomadic hunter-gatherers, lived in small villages on the wooded slopes of Mount Pinatubo. Within the mountain, according to their beliefs, lived their supreme god, Apo Mallari. But in the spring of 1991, their god grew restless, and in June of that year Apo Mallari awoke fully, belching volcanic ash into the stratosphere in the second-largest volcanic eruption of the twentieth century. Fortunately, most of the Aeta had been evacuated or had simply fled before the worst of the eruption. For many, however, it meant permanent displacement. Their homes and their livelihoods destroyed, they were dispersed to the lowlands, some to "temporary" camps that turned into long-term living quarters. It was a near-fatal blow for these indigenous people, whose existence as a distinct group was already precarious.

The volcano was not selective in its destruction. Less than ten miles to the east of Mount Pinatubo lived another group of people, inhabitants of the Clark Air Force Base, the largest U.S. military base outside America, complete with schools, a shopping mall, a cinema, and much else. Like the land around it, the base was blanketed with ash from the volcano. During one eruptive burst, a high-speed avalanche of glowing ash and dust—in geological terms, a pyroclastic flow—roared down the mountain toward the base and stopped just short of engulfing it. Ash choked the motors of aircraft and ground vehicles, and its weight collapsed the roofs of aircraft hangars. Within months the entire huge facility had been abandoned and would never reopen.

Mount Pinatubo is just one of many active volcanoes in the Philippines. Its eruption in 1991 was awe-inspiring even to those who saw only pictures or video footage; for those who witnessed it firsthand, it was, in the true sense of the phrase, mind-boggling. But the eruption was not particularly unusual in geological terms. On the VEI (volcanic explosivity index) scale, which measures the intensity of eruptions, it ranks as a 6. There were three known eruptions in this category during the twentieth century. (This is higher than the long-term average; over the past ten thousand years, there has been only one such eruption every few hundred years.) Like the earthquake magnitude scale, the VEI is open-ended, with each unit increase on the scale denoting a factor-of-ten increase in intensity. But eruption size is not measured solely by the response of an instrument, as is the case for earthquakes. The assessment is more qualitative, based on estimates of quantities such as the amount of material ejected, the height of the eruption plume, and the duration of eruptive activity. For unobserved eruptions of the past, generally the only measure possible is the amount of material ejected, which can be determined from the thickness of volcanic ash layers and lava flows and their geographical extent.

Mount Pinatubo is a subduction zone volcano, part of the so-called ring of fire that surrounds most of the Pacific Ocean. Like most subduction zone volcanoes, it erupts explosively, in dramatic fashion, spewing gas, solid rocks, and blobs of molten magma into the air at high velocities. The power for this explosive activity comes from water and other volatile compounds dissolved in the magma. As molten rock approaches the surface, the gases bubble out of the magma and expand rapidly. The chemical composition of subduction zone magmas enhances this behavior: they are rich in silica, which makes them very viscous. The expanding gas bubbles are constrained in the viscous lava until very high pressures build up in the volcanic conduits, and violent, explosive eruptions are the result.

The 1991 eruption of Mount Pinatubo broke almost five hundred years of quiet. Prior to the eruption, the mountain did not have the appearance of an active volcano. There were a few hot springs, but no fresh lava flows, and its flanks were heavily eroded and thickly forested. But in mid-March of 1991, a series of earthquakes shook the mountain, and over the next few weeks more and larger earthquakes were felt. It soon became clear that the volcano was awakening from a long slumber; the earthquakes were a sign that magma was moving toward the surface. Then, in early April, several small eruptions of superheated steam occurred near the summit, sending debris clouds into the air and coating the surrounding countryside with a thin veneer of volcanic dust.

The earthquakes and small explosions were disturbing, but to a casual observer they didn't seem too serious because they were not accompanied by eruptions of fresh lava at the summit. But the steam explosions were a sign that there was an intense source of heat near the surface, superheating groundwater until steam exploded upward along cracks and fractures. Geologists at the Philippine Institute of Volcanology and Seismology and their colleagues at the U.S. Geological Survey recognized that an eruption might be imminent, and they quickly set up a network of instruments on and around the volcano: seismometers to accurately measure the size and location of the small earthquakes that tracked the rise of magma, and tiltmeters to monitor the changing shape of the mountain as it inflated in response to upwelling molten rock.

Mount Pinatubo lies in a heavily populated part of the world. In 1991 there were hundreds of thousands of people living within a few dozen miles of the volcano—a distance well within the danger zone for a large, explosive eruption. Manila, only fifty-five miles away, is home to more than nineteen million people. The scientists monitoring the mountain were under pressure to get it right—it is easy to say "better safe than sorry," but a large-scale evacuation based on a false alarm would seriously undermine the effectiveness of future warnings.

All the signs pointed to an impending eruption, however. Mapping and dating of volcanic rocks in the area showed that Mount Pinatubo had experienced at least three very large explosive eruptions in the previous six thousand years. It also showed that the flatlands surrounding the mountain were floored by huge mudflows that had carried vast quantities of ash from the past eruptions down the mountainside into areas that were now heavily populated. An eruption on a similar scale would wreak havoc.

By early April 1991, the continuing earthquakes and steam eruptions had prompted the authorities to issue an evacuation order for people living within a few miles of the volcano. The activity continued throughout April and May, and although no lava erupted, scientists monitoring the volcano detected emissions of large amounts—up to thousands of tons per day—of the deadly volcanic gas sulfur dioxide. On June 7 magma finally reached the surface—not in an explosive eruption, but as a thick, slow-moving mass of lava that formed a dome at the top of the mountain (the gases that had been dissolved in this first batch of magma had apparently leaked out during the liquid's slow ascent to the surface, allowing it to erupt in a relatively nonviolent way). The Philippine Institute of Volcanology and Seismicity issued a level 4 warning, the second-highest alert on their scale of five. Just a few days later, fresh, gas-charged magma reached the surface, and explosive eruptions began. The warning level was raised to 5. Volcanologists continued to monitor the volcano closely, but they had no way to predict exactly what would happen next.

Then, on June 15, there was a truly enormous eruption that ejected more than a cubic mile of material from the volcano. Glowing pyroclastic flows—high-speed mixtures of red-hot gas and lava such as the one that almost engulfed Clark Air Force Base—raced down gullies and valleys on the sides of the volcano, filling them with a blanket of hot material hundreds of feet thick. (Five years after the eruption, measurements showed that the interiors of some of the flows were still as hot as 900 degrees Fahrenheit.) Some of the ejected material rose high into the atmosphere as fine ash and eventually spread around the entire globe. To add to the misery of those living close to the volcano, a typhoon struck just as the gigantic eruption occurred. The heavy rains washed volcanic ash out of the air and gave the tropical island of Luzon the appearance of a landscape emerging from a snowstorm. The wet ash was so heavy that it collapsed roofs indiscriminately, and for years after the eruption, ash and loose volcanic fragments surged down the mountainside in massive mudflows during the rainy season.

More than eight hundred people died as a result of the Mount Pinatubo eruption. Far more would have perished, however, without the prediction efforts that were hastily mounted once the volcano started to show signs of unrest. Regular bulletins and updates were issued via radio, TV, newspapers, and local organizations, so that almost everyone within the volcano's range was aware of the danger. Thousands were evacuated or advised to leave nearby areas. There is no way to stop a volcano from erupting, and there are no effective methods to mitigate the enormous physical damage caused by explosive eruptions, but the steps taken before and during the Mount Pinatubo eruption show that timely warnings based on careful monitoring can greatly reduce the human toll.

Alongside the devastation it caused on Luzon, the Pinatubo eruption also affected the environment globally. It was the first really large eruption for which such effects could be tracked by satellites and airborne monitors. The eruption blasted a plume of dust and volcanic gases more than twenty miles high into the atmosphere. Although fine volcanic dust settles out relatively quickly, one of the volcanic gases, sulfur dioxide, reacts with water to form aerosols in the stratosphere, tiny droplets of sulfuric acid that persist for years. The magma at Pinatubo was especially rich in sulfur dioxide, and the abundant aerosol particles that formed after the eruption partially blocked incoming sunlight. Global temperatures dropped by almost a degree Fahrenheit for more than a year following the eruption, temporarily slowing the march of global warming. The sulfur also enhanced ozone destruction in the stratosphere, causing the largest Southern Hemisphere ozone hole that had been observed until then. And although it was no consolation for the people of the Philippines, the aerosols and volcanic dust had one positive effect: they produced spectacular sunsets that were enjoyed by people around the world.

Explosive subduction zone volcanism like that at Mount Pinatubo typically results in picturesque, snow-capped, conical volcanoes known as "stratovolcanoes," like Mount Fuji in Japan. Large eruptions from these volcanoes often leave a crater in their wake, or, in geological terms, a caldera. Active volcanoes inflate and swell as upwelling magma nears the surface, but when large quantities of liquid rock are expelled quickly during the eruption, the upper part of the volcano simply falls in on itself like a punctured balloon. Frequently lakes form within these calderas, as has happened at Mount Pinatubo.

In the San Juan Mountains of the western United States, there is a caldera so huge that it took geologists several decades of exploration and mapping to work out its dimensions. One reason the task was so difficult is that this particular volcanic crater is 28 million years old and heavily eroded. It has also been partly filled in by later volcanic eruptions. The caldera is called La Garita, after the nearby Colorado town of that name. It is oblong in shape, measuring about forty-seven miles by twenty-two miles, and has the distinction of having been formed in the largest explosive eruption known in the geological record. The amount of material ejected from La Garita was enormous, more than 1,200 cubic miles of ash and lava with an estimated weight of more than eighteen trillion tons. By way of comparison, the Empire State Building weighs a paltry 365 *thousand* tons. The La Garita eruption ejected more than a thousand times as much ash and lava as the 1991 eruption at Mount Pinatubo, and it spread ash across much of the United States. Geoscientists who study explosive volcanism believe that La Garita was close to the physical size limit for this type of eruption on Earth.

Nobody was around to observe the La Garita eruption, so we don't know how long it lasted. But its main phase must have been short, probably lasting days to a few weeks, because its primary product—a deposit known as the Fish Canyon Tuff (*tuff* is a geological term for rock made up of volcanic ash and fragments)—is incredibly uniform. It has the same chemical composition and contains the same assemblage of minerals everywhere it is found, and is classified by geologists as a "single cooling unit," meaning that as far as can be discerned, the entire deposit—which is three-quarters of a mile thick in at least one area—is the product of a single eruption. The homogeneity of this enormous volume of material indicates that magma in the chamber underlying the volcano must have been very well mixed before it was erupted. Crystals of the potassium-rich mineral feldspar from the tuff are widely used as a standard for potassium-argon dating; they date the eruption very precisely to 28.2 million years ago, plus or minus 46,000 years.

The environmental consequences of the La Garita eruption were severe. We don't know how much ash or sulfur dioxide was thrown into the atmosphere, but the sheer size of the eruption means that the effects would have been orders of magnitude greater than those of Pinatubo, and it is likely that global temperatures were depressed, perhaps by tens of degrees, for an extended period. Large parts of the western United States would have been plunged into darkness for the duration of the eruption and for days or weeks afterward, and much of the country would have been blanketed by suffocating layers of ash. Surprisingly, perhaps, the fossil record shows no obvious rise in extinction rates 28.2 million years ago.

La Garita may not be linked directly to global extinctions, but it certainly would have had deadly consequences regionally. Even much smaller explosive eruptions can have devastating effects on life over a wide area, as was discovered in 1971 by paleontologist Mike Voorhies. Voorhies was searching for fossils in northeastern Nebraska, a region that was home to extensive grasslands and teeming with animal life more than ten million years ago. At the side of a ravine, he came upon the skull of a baby rhinoceros, poking out of a layer of soft, eroding volcanic ash. Fossil rhinoceroses are fairly common in the area, but what made this one unusual is that all the bones of the skull and jaw were still properly connected together, not scattered and disjointed as they usually are in such finds. As he carefully brushed away the ash, Voorhies was astounded to find that not only was the head complete, but the whole animal was preserved as an undisturbed, three-dimensional skeleton.

This was a once-in-a-lifetime discovery. The baby rhinoceros, it turned out, was just one of hundreds of perfectly preserved fossils of now-extinct large animals entombed in the volcanic ash. In June 1991, twenty years after Voorhies found the rhinoceros skull and at almost exactly the same time that Mount Pinatubo was beginning to blast *its* volcanic ash into the atmosphere, the fossil location in Nebraska became the Ashfall Fossil Beds State Historical Park. Today it is both a tourist attraction and a working paleontological site. The park is the only place in the world where intact, three-dimensional fossils of large animals like rhinoceroses are found (see figure 28). Paleontologists have also uncovered fossils of camels, deer, several species of horses, and dogs, as well as many smaller animals, such as birds and turtles. All of these animals were killed and preserved because of a large explosive volcanic eruption that occurred almost a thousand miles to the west.

In order to understand the geological context of the unique Ashfall Park fossils, the ash has been carefully mapped and analyzed. The prevailing sedimentary rock in the region is sandstone, and the ash in which the fossils are preserved forms a distinct layer within the sandstone. The thickness of the ash varies considerably from place to place, from a few feet to more than six feet, and the mapping shows that it is thickest in what were originally depressions and low-lying ground, probably because it was blown about in the wind like drifting snow. The ash is made up almost entirely of delicate shards of volcanic glass, the broken walls of expanding gas bubbles in the erupting magma that froze instantly to form natural glass when they hit the cold air at the Earth's surface. Through its chemical characteristics, the glass can be matched to a large caldera in southeastern Idaho, formed in an explosive eruption about twelve million years ago.

The prevailing westerly winds would have carried ash the thousand

Figure 28. Morris and McGrew, two adult barrel-bodied rhinoceroses (*Teleoceras major*) unearthed from volcanic ash at Ashfall Fossil Beds State Historical Park, Nebraska. (Courtesy University of Nebraska State Museum.)

miles from the volcano to the Nebraska site within little more than a day, and it would have begun to affect the local wildlife immediately. The rhinoceros found by Voorhies, and most of the other well-preserved fossils, come from deposits laid down in a waterhole, evidently one frequented by many animals. There is a hierarchy in the preservation of the fossils—the smallest animals succumbed first and are at the bottom; the larger ones are on top. Some of the larger animals still have the remains of their last meals—mostly prairie grass—in their jaws and abdomens. They also show signs of lung disease. Breathing the fine volcanic ash, and probably also ingesting it as they munched on grass or drank from ash-choked ponds, was lethal and killed them within days. It is a cautionary tale—ecosystems all along the thousand-mile path of the ash cloud, and most likely far beyond, must have suffered similar fates. The eruption responsible for the mayhem was larger than the 1991 Pinatubo eruption, but it was much smaller than La Garita.

The very largest eruptions known, with La Garita heading the list, are now sometimes referred to as "supervolcanoes" or "supereruptions." The term *supervolcano* is a recent designation; it was coined not by scientists, but in a TV documentary about large explosive eruptions, and is now in widespread use. Although, strictly speaking, "supervolcano" and "supereruption" refer to different things, the words are often used interchangeably. Formally, they refer to eruptions ranked 8 or higher on the VEI scale (meaning they are at least a hundred times bigger than the Pinatubo eruption). In terms of the volume of erupted material, to qualify as a VEI 8 supereruption, a volcano must eject a volume of 240 cubic miles (1,000 cubic kilometers) or more of material.

Undoubtedly, there have been supereruptions throughout the Earth's history, but nearly all of the known examples date from the past fifty million years or so. Erosion, plate tectonics, and sometimes burial under younger sediments have obliterated calderas and other telltale clues to the existence of most older ones, although a few can be inferred from widespread ash layers preserved in sedimentary rocks.

In a recently compiled list of the largest known explosive volcanoes, significantly more than half are located in the western United States. Partly, this may be due to the fact that vegetation cover is sparse and the region has been well scrutinized by geologists. But the geological setting also plays an important role. For most of the past 300 million years, a subduction zone ran along the western edge of North America. Arcs of volcanoes developed above it, and bits of continents and island arcs were plastered onto the margin of the continent, extending it westward and making geological maps of western North America look like patchwork quilts. Subduction provided the raw material for some of western North America's supervolcanoes by dragging water-laden sediments and ocean crust down under the edge of the continent, releasing volatiles that fluxed the mantle below and induced melting. Some of the explosive volcanism in western North America, like that at La Garita, occurred far inland, a long way from the subduction zone, but even in these cases it was probably linked to the effects of subduction on the crust and the underlying mantle at the continent's edge.

After La Garita, the next-largest known supereruption occurred at Mount Toba, in Indonesia, about 74,000 years ago. Mount Toba is unequivocally related to a subduction zone, sitting directly above the region where the north-moving Indian plate slips under the island of Sumatra. The great Indonesian earthquake of December 26, 2004, which generated a large tsunami throughout the Indian Ocean, took place along the same subduction zone. Mount Toba has experienced many violent eruptions in the past, as is evident from three large, overlapping calderas near its summit; the largest (and most recent) of these was formed during the event 74,000 years ago. The total amount of material ejected was about 675 cubic miles, a bit more than half that estimated to have been erupted at La Garita (see figure 29). Volcanic ash from the Toba eruption covered all of India and much of South East Asia and can be traced from the Arabian Sea in the west to the South China Sea in the east; in total it spread over nearly 4 percent of the Earth's surface.

The Mount Toba eruption has drawn attention because of its enormous size, but its timing is also important: 74,000 years ago modern humans already populated the Earth. Did the environmental effects of this eruption affect our ancestors? In 1998 Stanley Ambrose of the University of Illinois made a startling proposal. The Toba eruption, he said, might have caused a "bottleneck" in human evolution that had previously been identified by geneticists. Population bottlenecks are well known in many animal species; they are recognized when genetic evidence shows that all current members of a species trace back to a small number of ancestors at some specific time in the past. Some bottlenecks have been historically documented; for example, there are cases of endangered species having come close to extinction and then rebounding. The prehistoric human bottleneck, which appears to have occurred between about 50,000 and 100,000 years ago, saw a severe drop in population size, probably to ten thousand individuals or perhaps fewer, followed by expansion toward the present-day human popula-

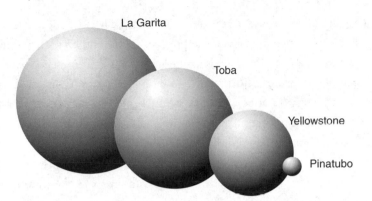

Figure 29. A comparison of estimated eruption volumes for the volcanoes discussed in this chapter. The actual volumes range from about 1,200 cubic miles (about 5,000 cubic kilometers) for La Garita to approximately 2.6 cubic miles (about 11 cubic kilometers) for Pinatubo.

tion. Ambrose proposed that the environmental effects of Toba, particularly the sharp temperature drop caused when sulfuric acid aerosols dimmed the Sun, resulted in widespread famine and decimated the existing human population except in a few tropical refuges in Africa.

Ambrose's theory is controversial. In particular, recent archeological work at a site in India—not so very far from Toba—has unearthed evidence that at least one human group survived the eruption relatively unscathed. The excavations revealed that the same kinds of tools occur both beneath and above a volcanic ash layer that can be conclusively linked to the Toba eruption, indicating that they were in use both before and after the event. No fossils have been found with the tools, so nothing is known about the people who made and used them, but whoever these ancient people were, they went on living in much the same way after ash from Toba blanketed their village as they had before. The evidence tells us nothing about their numbers, however, and it is conceivable that only a small fraction of the original population survived the eruption.

The fate of Ambrose's idea is thus unclear; it may turn out to be completely wrong. Certainly, it is not supported by the Indian archeological data, and there are also questions among biologists about the timing and size of the "genetic bottleneck" in human evolution. But this story is a neat illustration of how science works: Ambrose recognized a possible coincidence between a natural catastrophe and a feature of human evolution, and hypothesized that there might be a connection. He mustered evidence (timing, expected environmental effects of the eruption, and so on) for his idea, and when it had its moment in the limelight it prompted biologists and archeologists to look closely at other data that could test the hypothesis. Whether the idea survives this scrutiny or not, the research done to explore it will almost certainly lead to a better understanding of Toba's effects on humans.

Whatever the outcome of that work, the catastrophic nature of the Toba eruption cannot be denied. In ice cores from Greenland, for example, almost halfway around the Earth from Sumatra, the largest sulfur pulse in more than 100,000 years occurs in ice layers deposited at the time of the Toba eruption. This finding indicates that huge quantities of sulfur dioxide were emitted from the volcano; the best current estimates put the amount at more than one hundred times that from Mount Pinatubo in 1991. The closely monitored aerosol cloud from Pinatubo circled the Earth within a few weeks and spread globally within months; as noted earlier, it lowered global temperatures by about a degree Fahrenheit for more than a year. Climate models indicate that aerosols from the Toba eruption caused global average temperatures to drop by as much as eighteen degrees Fahrenheit for a period of several years. Temperatures recovered to near their pre-eruption values within about a decade, but even such a relatively short interval of global cold, coupled with decreased precipitation (also predicted by the models) may have killed off large tracts of tropical rain forests. The significantly lower levels of sunlight caused by the aerosol particles would have exacerbated other environmental stresses.

No supereruption has occurred in recorded history, so the expected environmental effects must be extrapolated from smaller, observed eruptions such as Pinatubo. However, all current evidence suggests that this is a reasonable approach—there is nothing to suggest that super-eruptions are qualitatively different from their smaller cousins. If the model results for Toba are correct, ecosystems worldwide would have been severely stressed, even if the environmental effects were short-lived. High levels of dust occur in the Greenland ice cores shortly after the Toba sulfur pulse, possibly a sign that extensive vegetation dieback exposed large tracts of land surface to wind erosion.

Conclusions about the environmental aftermath of the Toba eruption find support from the historically documented effects of another large eruption in the same region of the world. In April 1815 Mount Tambora—located, like Mount Toba, in Indonesia—erupted explosively and caused widespread devastation. The eruption was too small to be classified as a supereruption, and the reports of its effects are qualitative rather than quantitative. Even so, it is quite clear that both its local and its global consequences were severe. Tens of thousands of people died either from the immediate effects of the eruption or from the crop loss and famine that followed. In the Northern Hemisphere—the location of most of the world's population and agricultural land—the year following the Tambora eruption came to be known as the "year without a summer." Snow fell in Canada and the United States in June and July, and on both sides of the Atlantic crops and livestock died off, precipitating the worst famine of the nineteenth century. Unusually low temperatures continued for the next two years. If the Tambora eruption precipitated such dramatic change, the global consequences from Toba—fifty times larger—must have been catastrophic.

The supereruptions at Mount Toba, and probably at La Garita, were directly connected with subduction zone processes, but this is not the case for all supervolcanoes. Although most people don't realize it, there is an active supervolcano sitting right on the doorstep of many Americans, related not to subduction, but to a mantle hotspot. The volcano's name? Yellowstone.

To most of us, Yellowstone is a tourist destination; it conjures up images of spectacular scenery and Old Faithful blowing off steam. But the geysers and hot springs that make Yellowstone Park the world's largest concentration of such features are only the surface signs of a buried reservoir of red-hot magma not far below. Geological investigations of the Yellowstone region indicate that over the past several million years the magma has periodically exploded in gigantic eruptions, the most recent of which occurred 640,000 years ago. That eruption spread ash over much of the central and western United States and into Mexico and Canada, and created the huge Yellowstone caldera, the central geological feature of the park today, fifty miles long and thirty-five miles wide. An obvious question is, Will this restless giant awaken again any time soon? It is very unlikely, but look far into the future—hundreds of thousands of years or more—and it is inevitable. The nature of Yellowstone's volcanic activity makes it virtually certain that sometime in the distant future another devastating eruption will occur. That knowledge alone has sparked public and scientific interest in Yellowstone volcanism. It was even the subject of a documentary film aired by the BBC in 2005.

Like all other supervolcano eruptions, those at Yellowstone required the accumulation of vast quantities of viscous molten rock, charged with water and other volatiles, in shallow-level reservoirs in the Earth's crust. The source of the magma at Yellowstone is a rising plume of hot material originating deep in the Earth's mantle. Just as the Hawaiian chain of volcanoes traces the path of the Pacific plate over a fixed mantle hotspot, a series of calderas extending southwest from Yellowstone across Idaho to the Oregon-Nevada border traces the movement of the North American plate over the Yellowstone hotspot (see figure 30). Radiometric dating shows that these calderas become progressively older the further away they are from Yellowstone, with the most distant dating to sixteen million years ago. The path of the Yellowstone hotspot across the landscape is marked by the Snake River Plain, a relatively flat swath roughly sixty miles wide that cuts through the locally

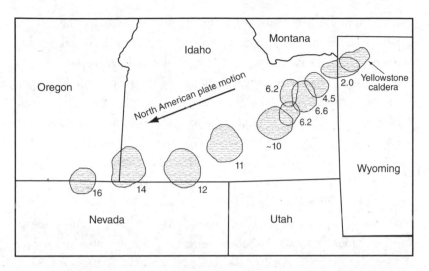

Figure 30. A series of large calderas across the western United States traces the movement of the North American plate over the Yellowstone hotspot, which has remained approximately fixed while the plate has moved westward over it. Approximate ages of the caldera-forming eruptions are given in millions of years (some are not known with certainty). The Yellowstone caldera formed 640,000 years ago.

mountainous terrain. The plain was formed by lavas and ash from the giant caldera-forming eruptions and associated volcanic activity, which completely buried local topographic features. Ash from one of these eruptive centers was responsible for burying the rhinoceroses found at Nebraska's Ashfall Fossil Beds State Historical Park. Today the hotspot sits directly beneath Yellowstone; seismic imaging has traced the narrow conduit of hot material deep into the mantle.

Magma produced by mantle hotspots penetrates the oceanic crust relatively easily to form volcanoes like those of Iceland or the Hawaiian Islands, but it is generally too dense to rise through the thicker and much lighter continental crust. It does, however, supply enough heat to melt parts of the crust, and it is this process that generates most of the magma ultimately erupted from hotspot-linked supervolcanoes. Also, in a process roughly akin to distillation, water and other volatile compounds escape upward into the shallow magma reservoirs lying under the volcanoes, and these pressurized gases supply the energy for explosive eruption.

The time between massive eruptions over the Yellowstone hotspot, reflected in the ages of the calderas they have produced, provides a clue to the time it takes for enough magma and gas to accumulate before a supereruption occurs. The data in figure 30 show clearly that a few million years are typically required, but also that the process must be sporadic rather than continuous because the time gaps between eruptions are variable. There is also evidence that extensive but less violent volcanic activity has occurred between each of the caldera-forming supereruptions. At Yellowstone, for example, intermittent lava flows have filled up much of the Yellowstone caldera since it formed 640,000 years ago; the youngest of these flows, which dominate the present-day landscape, dates to 72,000 years ago. There are no signs that renewed volcanic activity is imminent at Yellowstone, but because *any* eruption would be locally damaging, and a giant explosive eruption would have far-reaching consequences for the entire United States and indeed the world, an intensive monitoring program has been mounted to determine the current state of magma accumulation beneath the caldera. The responsibility for coordinating this work rests with the Yellowstone Volcano Observatory (YVO), a joint venture of the U.S. Geological Survey, the University of Utah, and Yellowstone Park. You can read regular monthly updates about Yellowstone earthquake activity, ground movement, and other related topics on the YVO Web site.

If there was ever any doubt that Yellowstone is still volcanically active, it has been dispelled by the monitoring results. Seismographs record several thousand earthquakes every year, GPS and other technologies record cycles of uplift and subsidence throughout the caldera at rates up to several inches per year, the flow of heat from the ground is about thirty times greater than the regional average, and there is a huge emission of volcanic gases (dominated by an estimated 4,500 tons *per day* of carbon dioxide). The primary challenge facing scientists involved in the monitoring effort is interpretation. Although the results have clarified much of what is happening in the Earth's crust under the caldera, it is not yet known which signals are most likely to be precursors to the next volcanic eruption.

Remote sensing data from the geophysical equivalent of a CAT scan show that there are distinct hot regions in the crust beneath the Yellowstone caldera, probably partly filled with magma, at depths between about five and twelve miles. The locations of small earthquakes and the up-and-down ground movements observed at Yellowstone suggest that the magma bodies are not static, but migrate within the crust. An important goal for those monitoring Yellowstone's activity is to determine whether these magma movements, and the high heat flow and carbon dioxide emissions in the region, result from injection of new magma from the hotspot far below, or whether they can be attributed to slow cooling and crystallization of existing magma beneath the caldera. This question has not yet been answered, but it is a crucial one, because it could indicate whether or not the volcano is building up toward a new eruption.

Regardless of where Yellowstone sits in its cycle of magma accumulation, however, the remote sensing data indicate that the amount currently stored below the caldera is too small to feed a catastrophic eruption, so that possibility can be pushed far into the future. The most probable near-term hazard at Yellowstone comes not from a true volcanic eruption, but from a steam explosion. Although these have been very common in the history of the caldera and are well studied, exactly what triggers them is unclear, making prediction difficult. At least some steam explosions appear to occur when there is a change in the "plumbing" system that circulates water deep underground and supplies the hot water and steam for geysers and hot springs. If an earthquake suddenly opens new fractures to the surface, for example, and releases the pressure on deep, superheated water, the water instantly expands into steam and explodes upward through the fractures. Steam explosions that throw out rocks and mud and create small potholes filled with hot water are relatively frequent at Yellowstone; at least one is documented every few years and many more probably go unnoticed because the park is large and sparsely populated. Large steam explosions, although rare, also occur at Yellowstone, and these can be very destructive. The largest, dated to fourteen thousand years ago, created a crater one and a half miles in diameter—bigger than the Barringer meteorite crater.

There are a number of potentially active supervolcanoes like Yellowstone on Earth, and the recognition of just how globally disastrous an eruption from one of them would be has made supereruptions a topic of worldwide concern among geoscientists. The geological record shows that supereruptions occur significantly more frequently than collisions with small asteroids that could cause equivalent damage, which makes supervolcanoes one of the most serious long-term geological threats to mankind. And while it may eventually be feasible to divert an incoming asteroid, there is at present no way to halt or even mitigate a supereruption. In fact, there are not even any remotely plausible ideas for doing so. Until there are, close monitoring of "active" supervolcanoes like Yellowstone, and preparation for the difficult aftermath of a giant eruption, are the only feasible options.

Even that, however, is easier said than done. Almost certainly, many volcanoes that have the potential to erupt explosively in large, damaging eruptions have not yet been recognized, let alone monitored. Scientific organizations in the United States and the United Kingdom, as well as an international group of volcano researchers, the International Association of Volcanology and Chemistry of the Earth's Interior, have set up working groups and issued reports in an attempt to make recently acquired knowledge about large explosive eruptions more widely known. In the United States, the Geological Survey is at the forefront of monitoring potential eruption sites. Hopefully, these efforts will succeed in raising awareness of the hazards posed by very large explosive eruptions before the next restless giant awakes. As the experience of the 1991 Mount Pinatubo eruption shows, advance warning is the key to minimizing human suffering from volcanic eruptions, whatever their size.

CHAPTER TWELVE

Swimming, Crawling, and Flying toward the Present

In 1818 Adam Sedgwick was named the Woodwardian Professor of Geology at the University of Cambridge in England. The chair had been endowed by a famous predecessor almost a century earlier, and the appointment was a great honor for Sedgwick. But he had no formal training in geology. Legend has it that he quipped, "Hitherto I have never turned a stone; henceforth I will leave no stone unturned." Whether or not he really did utter those words, Sedgwick lived up to them. He became one of the nineteenth century's foremost proponents of the relatively new science of geology, inspiring students at Cambridge with his enthusiastic and articulate lectures. One of those students was a young Charles Darwin, whom Sedgwick befriended and took on as an assistant for a summer field excursion in Wales. The two remained friends, and Darwin sent Sedgwick fossils, rocks, and geological notes as he traveled the world on the *Beagle*. Some of those samples are still in Cambridge—in the university's Sedgwick Museum of Earth Sciences.

Sedgwick's work in Wales, where Darwin cut his teeth on geology, provided the framework for the first geological period of the Phanerozoic eon, the Cambrian period. Sedgwick called the rocks he worked on in Wales the "Cambrian System" after the Roman name for Wales, Cambria. He recognized that the Cambrian rocks were very old, but in a few places he found other rocks, lying beneath the Cambrian strata, that were even more ancient. While the Cambrian layers contained fossils, the older rocks were barren of any signs of life. Because of this, Sedgwick and many of his geological colleagues believed that the Cambrian rocks recorded the first arrival of life on Earth.

As is evident from earlier chapters in this book, we now know better. Without question, the Phanerozoic eon, beginning with the Cambrian period, witnessed an unprecedented proliferation and diversification of life until it eventually occupied every conceivable ecological niche in the oceans, on land, and in the air. The initial burst of evolution has been called the "Cambrian explosion" because of its rapidity and broad reach. But when the Cambrian period began 542 million years ago, life already had a history that stretched back billions of years. The difference, for early geologists, was that Cambrian animals were the first to develop hard body parts like shells and protective chitinous armor: they were much more easily preserved than their soft-bodied predecessors, and as a result Cambrian and later sedimentary rocks usually contain easily recognized fossils. It is not difficult to understand why Sedgwick and many other geologists thought that life first appeared in the Cambrian.

Several of the important events of the Phanerozoic eon have already been discussed in this book, but to round out our overview of the Earth's history the current chapter takes a broad look at how the planet, and especially how life on it, has evolved over the most recent 20 percent of its history. The closer we get to the present, the more we know—something that is evident from the timescale shown in figure 1. Abundant clues in the rocks mean that the Phanerozoic can be much more finely subdivided than earlier parts of the Earth's history.

The Earth at the beginning of the Phanerozoic eon was still a very different planet from the one we now know. The shapes of the continents bore little resemblance to those of today, and their arrangement on the globe was also unfamiliar (see figure 33 on page 242). Most landmasses were located in the Southern Hemisphere. Laurentia, the continent that would eventually become North America, lay near the equator. The global ice ages of the Proterozoic eon, the Snowball Earth episodes, were a distant memory; carbon dioxide levels in the atmosphere were high, the climate was warm, and with little or no water locked up in permanent glacier ice, sea levels were high and significant portions of all the continents were submerged under shallow inland seas. Oxygen levels in the atmosphere were only slightly lower than those of today, allowing complex animals to flourish in the oceans. But the continents were still barren of plants, and without a protective cover of vegetation, they were subject to the harsh forces of erosion.

Figure 31 is a simple time line for the Phanerozoic, showing some of the more important events that have taken place during this slice of geological time. Notable on the diagram are the periodic mass extinctions that have occurred over the past half-billion years. Progression from the relatively simple plants and animals that existed at the beginning of the Phanerozoic eon to the complex array of life that populates the Earth today has been a bumpy road, with occasional mass dieoffs along the way. While extinctions of individual species are a normal part of the evolution of life, multiple plant and animal groups were simultaneously affected during these rare, short intervals of mass extinction, and during some of them species disappeared hundreds or even thousands of times faster than usual. As we have seen in previous chapters, the mass extinction intervals appear to have been caused by unusual environmental conditions, although there is still no universal agreement about just what those conditions were in each case.

Long before the term *mass extinction* was coined, the existence of these events was implicitly recognized in the structure of the geological timescale. Boundaries were placed where the fossil record shows large and abrupt changes; two of the most severe mass extinctions on record mark the boundaries between the major divisions (the eras) of the Phanerozoic eon. Those extinctions, at the Permian-Triassic (P-T) and Cretaceous-Tertiary (K-T) boundaries, are two of what have come to be known as the "big five" Phanerozoic mass extinctions.

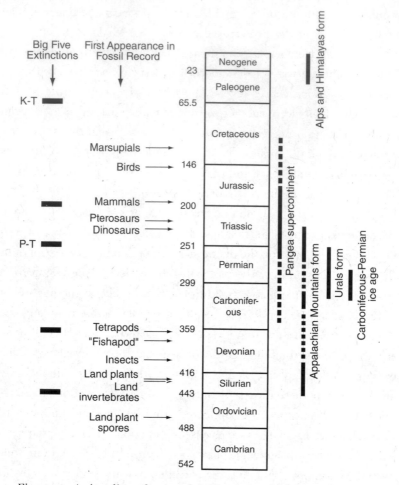

Figure 31. A time line of some of the important biological and physical events of the Phanerozoic eon. Dates are given in millions of years before the present. The Pleistocene Ice Age of the past few million years is not shown.

The physical environmental changes that have been suggested as triggers for these extinctions include, among others, global cooling and glaciation, the environmental effects of extensive volcanic activity (including global warming caused by release of greenhouse gases), acidification of the oceans due to high carbon dioxide levels, the release of toxic hydrogen sulfide gas from an anoxic ocean, and the multiple effects of an impact. In recent years another cause of mass extinction has been recognized: human activity. Most biologists believe we are currently in the midst of a biodiversity crisis that may result in 50 percent or more of all existing species dying out within the next century. That is extinction at an unprecedented rate, even in comparison to the most extreme estimates for the "mother of all extinctions" at the P-T boundary. The current crisis has been called the "sixth extinction," and unless environmental changes are slowed, it may rival or even exceed the big five of the Phanerozoic. The causes of present-day extinctions—all ultimately traceable to human activity—are many, and include reduction of habitat, targeted exploitation such as hunting and (especially) commercial fishing, and global warming induced by fossil fuel burning.

Information about the big five extinctions of the Phanerozoic comes primarily from the fossil record of animals that lived in the oceans. These are the most abundant and biologically diverse fossils available, and they provide a global sample because marine fossils occur on every continent. There are also data for land-based life, especially for the K-T extinction, the most recent of the big five, but in general the land fossil record is very sparse compared to that from the oceans. Even for marine animals, however, only a small fraction of the species present at any given time is preserved; as pointed out in the first chapter of this book, the fraction is estimated to be about 1 percent, which means that we will never know about most animals that have lived in the sea. Nevertheless, even though many plants and animals are missing, the fossil record through the Phanerozoic provides a clear outline of evolution and the role mass extinctions have played in it.

Still, there has been much soul-searching among paleontologists about whether the mass extinctions might somehow be illusions, the result of inadequate sampling, or unrecognized gaps in the record, or bias introduced because geographically restricted biological crises have inadvertently been sampled rather than global ones. But various rigorous statistical tests have been carried out that negate such possibilities. When extinction rates are examined for short time slices through the entire Phanerozoic eon, the five great extinctions stand out as times of especially large reductions in biological diversity. Numbers of plant and animal species, genera, and families all decline sharply. Most analyses focus on the number of species or genera that disappear, and those statistics show that the change in genus diversity is especially large for three of the big five extinctions: at the P-T boundary, at the Ordovician-Silurian boundary 443 million years ago (see figure 32), and at the K-T boundary. Some paleontologists have argued that only these three should be classified as true mass extinctions.

The K-T mass extinction has received the most attention because it is the only one for which there is indisputable evidence of an instantaneous, catastrophic event coincident with the extinctions. But in terms of sheer numbers of species extinguished, it ranks only third. At the Permian-Triassic boundary, at least 95 percent of animals living in the oceans disappeared; at the Ordovician-Silurian boundary, the dieoff was about 85 percent. At the K-T, it was around 75 percent. Some paleontologists have suggested that because of its severity, the P-T extinction was a very close call—that life on Earth was nearly eliminated entirely. That may be a bit alarmist—bacteria and archaea, for example, had already made it through several billion years of the Earth's history that included Snowball Earth glaciations, supergreenhouse episodes, radical changes in ocean chemistry, and, almost certainly, asteroid impacts even more destructive than the one that would take place at the K-T boundary. These organisms could navigate through even harsher biological crises than the one at the P-T boundary.

Surprisingly, perhaps, the Ordovician-Silurian event, the most ancient of the Phanerozoic mass extinctions, is very well documented. As the

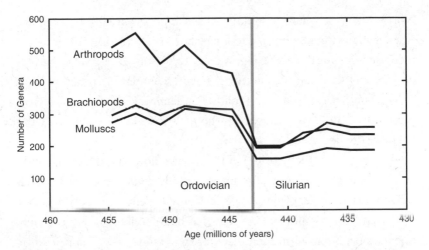

Figure 32. The diversity of various marine animal phyla, expressed as the number of genera found as fossils in two-million-year time intervals across the Ordovician-Silurian boundary. All of these animals show a sharp drop in numbers at the boundary, indicated by the dark vertical line. (Data from J.J. Sepkoski, A Compendium of Fossil Marine Genera, *Bulletins of American Paleontology* 363 [2002].)

Ordovician period drew to a close, the Sun's intensity was still about 5 percent lower than it is today, but high levels of carbon dioxide in the atmosphere—estimated to have been many times the present concentration—kept the greenhouse effect strong and the global climate warm. There were no extensive ice sheets on any of the continents. Today's southern landmasses were aggregated together into the large continent Gondwana, a remnant of the then still-fragmenting supercontinent of Rodinia. But as Gondwana drifted south into polar regions the climate cooled abruptly and an ice age began. Glacial deposits from the time are found in much of northern Africa and eastern South America—regions that were at or close to the South Pole.

What could have initiated an ice age at a time when greenhouse gases and global temperatures were so high? That is a vexing question; one suggestion is that mountain building, especially in eastern

North America, where mountains that would eventually become the Appalachians were beginning to form above a subduction zone, caused increased chemical weathering, rapidly depleting atmospheric carbon dioxide and lowering temperatures globally. As glaciers began to form at high southern latitudes on Gondwana, feedback effects, particularly the increase in albedo caused by the new ice sheets, accelerated the temperature decrease.

Whatever the precise origin of the ice age, however, the thick continental glaciers locked up a large amount of water, and sea levels dropped. This is clearly evident from the pattern of sedimentary rocks preserved from that time. Shallow inland seas that had flooded parts of the continents for a hundred million years or more suddenly disappeared. It was a double whammy for Ordovician animals: they were adapted to a warm climate and suffered when water temperatures decreased sharply, and at the same time the hospitable shallow-water environments where they thrived became rare because of receding seas. Furthermore, the changes were rapid, allowing little time for adaptation. Many species, genera, and families couldn't cope and became extinct.

In sedimentary rocks from pre-Phanerozoic times, clues to the Earth's surface environment come primarily from the rocks' chemical and physical characteristics. The appearance of fossils in the Cambrian period, however, provided an additional powerful indicator, because life is highly sensitive to environmental change. Individual species may be adapted to a specific narrow temperature range, or to shallow or deep water, or they may be constrained by aspects of seawater chemistry. Especially at times of mass extinction, the presence or absence of different types of fossils therefore serves as an illuminating record of changing environments.

At the Ordovician-Silurian transition, every animal group that has been studied suffered greatly, and the fine-scale detail of which species, genera, and families became extinct provides a good example of how successful geologists have become at using fossils to decode the past, even when the record is almost half a billion years old. Take, for example, the brachiopods—shelled marine animals that look a bit like clams. Brachiopods exist in today's oceans, but most live in deep, cold water or in high-latitude regions, and as a result they are not very common in shell collections. In Ordovician times, however, they were very abundant and occupied many different habitats, and their fossils are widespread.

The pattern of brachiopod extinctions at the Ordovician-Silurian boundary indicates that glaciation was a key factor in the biological crisis. During much of the Ordovician period, distinct brachiopod groups existed in different environments; they evolved over time, but as communities they changed little, each group occupying its own ecological niche. Two of these communities were adapted to life in warm, shallow water, one of them living in open ocean environments around the edges of continents, and the other occupying the inland seas that flooded the continents. These groups existed separately, with little overall change over long time periods and with no significant intermingling. But when rapid glacier growth drained the shallow inland seas, the brachiopod community that inhabited them was decimated. Their relatives along the continental margins also suffered, but not nearly to the same degree. The brachiopods living in inland seas apparently could not adapt or migrate quickly enough to survive, and they became extinct. When the ice age ended, however, the climate warmed, sea level rose because of the melting glaciers, and shallow seas again flooded the continents. Surviving brachiopods from the open-ocean, continental-margin community—a group that had never before invaded the inland seas—evolved, adapted, and quickly took over the newly re-created continental seas. These changes were direct responses to the environmental effects of glaciation—and they can be reconstructed in fine detail by carefully mapping out when and where the fossil organisms lived.

The course of the ice age can also be tracked through fossils of other brachiopod groups. The record shows that even many of those living at high latitudes, and thus already adapted to cool temperatures, were unable to cope with the rapidity and severity of the ice age cooling and became extinct. A few groups managed to survive by migrating to

lower, warmer latitudes, but then, when the ice age ended, they too succumbed as temperatures rose again. Detailed, layer-by-layer analysis
of sedimentary rocks shows that other animal groups also exhibit this
kind of double-dip extinction, with high extinction rates both at the
onset of the ice age and at its end. As is the case for all major extinctions, however, those groups that survived the periods of rapid environmental change enjoyed a spurt of evolution in the aftermath of the
crisis. Within a hundred million years, the brachiopods were even more
diverse than they had been before the extinction.

Each of the mass extinctions of the Phanerozoic eon has its own story
to tell, but in every case the fossil record shows that even many geographically widespread and successfully adapted species and biological groups became extinct. By definition, successfully adapted species
are able to withstand "normal" biological and physical stresses such as
predators, disease, and some degree of climate change, which indicates
that environmental disruptions during the mass extinctions were highly
unusual and probably abrupt, leaving little time for adaptation. Many
of the species that survived these crises did so because they happened
to possess characteristics that allowed them to withstand the rapidly
imposed environmental changes. It was indeed "survival of the fittest,"
but the required fitness depended on the nature of the change. This
essentially random process determined the future course of evolution;
the surviving groups expanded and evolved rapidly, taking over habitats
that had previously been occupied by other organisms. A classic example, discussed later in this chapter, is at the K-T boundary, where mammals—which were fairly common but not very significant before the
mass extinction—quickly replaced the dinosaurs in importance after it.

Although the big five mass extinctions had a disproportionate effect
on evolution, many of the most momentous changes in the nature of
life on Earth came about more slowly and in less precarious circumstances. One of the most important of these was the move from the
seas to dry land; even describing this transition as "momentous" is an
understatement. Just consider the consequences: when they moved onto

land, both plants and animals, accustomed to being supported in a fluid marine environment, had to develop completely new body structures to prevent collapse under their own weight. For plants, there was also the question of obtaining water: they were immobile, so how would they get water on dry land, and how would they retain it? Animals had to develop ways to obtain oxygen directly from air instead of from water. Both animals and plants had to contend with much bigger swings in temperature than they were accustomed to, on a daily and also on an annual basis. Animal eggs had to be protected from desiccation, and plant spores had to be distributed by some means other than flowing water. But the rewards were huge: beyond the oceans was a whole new world to populate.

Over the past few decades, there have been tremendous advances in our understanding of how the ocean-to-land transition occurred. Partly, this has been due to the search for and discovery of new fossils, often hailed in the media as "missing links" because they have filled gaps in the existing fossil record. Many of the new fossils have been discovered in the field, but some have also been unearthed from the dusty shelves of museums, where they had lain unrecognized and unidentified for years. The recent advances have also been fueled by new ways of looking at fossils: with X-rays, with electron microscopes, and with CAT scans and three-dimensional computer images that aid in understanding how extinct animals functioned.

Animals with spinal cords, our own distant vertebrate ancestors, are one part of this recently revised story. For many years, with few fossils to rely on and many gaps in the record, the conventional wisdom about how the vertebrates adapted to life on land was based on a theory by a famous Harvard University paleontologist, Alfred Romer. It was already well known that the earliest land vertebrates had evolved from fish, and that their limbs originated from fins. Romer proposed that creatures akin to today's lungfish—freshwater fish that can breathe air and are able to survive periods of desiccation by burying themselves in mud—evolved legs and eventually became fully land-dwelling as they pulled themselves from one water hole to another during times of drought.

Appealing as that image is, the anatomy of recently discovered fossils has turned the theory on its head. The adaptations necessary for life on land, it turns out, were developed by fish living in a fully aquatic setting, not by fish that somehow found themselves having to navigate on land. The adaptations began in shallow water environments along shorelines, possibly in partly brackish estuaries and marshes. Fins became stubby legs and feet, which were probably still useful as paddles for swimming but could also be used to push the animals along the bottom in shallow water. Skulls became flatter and disconnected from the shoulder, so that the animals could use their legs to push themselves up off the bottom and simultaneously raise their heads above the surface. Lungs developed alongside gills, and ribs grew to protect them. When the evolving animals raised their heads into the air, they could breathe. At this point they couldn't yet live on land, because their skeletons wouldn't support them. But they were getting there.

All land-living vertebrates are part of the animal group known as "tetrapods"—having four legs. This designation applies even to animals like birds, in which two of the legs have evolved into wings, and snakes, which obviously have no legs at all (the ancestors of snakes, though, did have legs, and vestiges of these ancestral limbs are still present in some kinds of snakes). In the fossil record, the very first tetrapods, still living in shallow water, date from about 363 million years ago. But in 2004 an even older fossil that seemed to be part fish and part tetrapod was found on Ellesmere Island in the Canadian Arctic. Its discoverers called it "Tiktaalik," an Inuit word meaning large fish. It has fins rather than limbs and is distinctly fishlike, but several of its other characteristics are those of tetrapods. When someone described it as a "fishapod," the nickname stuck. The rocks in which the fishapod was found are 375 million years old, and it clearly represents an important step in tetrapod evolution. Whether it is really the "missing link," or just one of a fascinating variety of sea creatures that were evolving toward life on land, is still debated.

Why did the tetrapods move onto land? "Because it was there" is one answer, and there may be some truth in it. We don't really know the details, but their ancestors may have ventured into shallow water to avoid predators, to feed on other fish that were seeking protection in vegetation-rich coastal estuaries and swamps, or simply—like turtles and crocodiles—to lounge in the sun. The transition onto land proceeded gradually. Even the first truly land-dwelling tetrapods, like immigrants to a new country, couldn't quite make up their minds about whether to stay or not: they were amphibians, returning to water to lay eggs or live out part of their life cycle.

The tetrapods were relative latecomers to the land. Plants had made the transition much earlier, as had small invertebrate animals such as the arthropods, a group that includes insects, crabs, and scorpions. Because fossils are much less likely to be preserved on land than in the oceans, there are many gaps in the record, and the oldest fossils of a particular land plant or animal group provides only a minimum estimate for their first appearance. These estimates are continually being pushed back as paleontologists discover new localities and fossils.

The first evidence that plants had moved onto land comes from spores found in sedimentary rocks from Oman. They have been dated at 475 million years, making them more than 100 million years older than the first land-dwelling tetrapods. What those early plants looked like, nobody knows, except that they bore little resemblance to the plants we are familiar with today. They were tiny, and their spores were surrounded by a protective cuticle, an adaptation to prevent them from drying out. Evolution soon produced such well-encased spores that they were almost indestructible. This was good for plant propagation, but also a great boon for paleontologists, because the spores didn't degrade. They were blown about by the wind and fossilized in the mud of lake bottoms and coastal seas, and they can be used to trace the evolution of the plants that produced them.

The earliest fossils of land plants, as opposed to their spores, are found in rocks that are about 425 million years old. Although they had already been colonizing the land for at least fifty million years, these plants were still very small and primitive, with no leaves and a very rudimentary root system. But fast-forward another fifty million years, and by then plants had evolved and developed tough, woody stems that held them upright and delivered water and nutrients to their branches and leaves. Large ferns and primitive trees dotted the landscape, especially in places where moisture was available, such as river deltas, estuaries, and along river valleys. This was the environment encountered by the first tetrapods when they emerged onto land from their swampy coastal habitats.

Small land animals also significantly predated the tetrapods. The first we know about was an inconspicuous millipede, and it has gained worldwide fame as the first air-breathing creature ever to walk on land. The millipede fossil was found in sedimentary rocks in Scotland dating from 428 million years ago, and is thus roughly the same age as the first land plant fossil. Although this creature has achieved celebrity status, there were certainly other and probably earlier invertebrate pioneers of dry land; they just haven't yet been found. In slightly younger rocks, dated between 396 and 407 million years old, researchers have discovered the remains of the first known true insect. The fossil is fragmentary, and although no wings were preserved, the scientists working on it found other characteristics indicating that it was a winged insect. Amazingly, using microscopes, the paleontologists were also able to observe and describe the fine details of this small, delicate insect's jaws and teeth. It was, they concluded from their analysis, a chewing insect. What exactly it crunched upon is not known, but the choice would have been limited. When the ancient insect was alive, plants were still small and had either tough, spiny leaves or no leaves at all. Plant spores, or perhaps other small invertebrates, might have been part of the early insect's diet. The important conclusion from these various fossil discoveries is that even more than 400 million years ago there was a diverse array of flora and fauna on land. The members of this group were small and their distribution was limited, but a foothold had been established. Fewer than

200 million years separate the first small shelly organisms of the early Cambrian seas from the fish, corals, land plants, insects, and amphibians of the late Devonian and early Carboniferous periods. Throughout the whole of the Phanerozoic eon, except for the brief intervals of mass extinction, there has been a steady increase in the variety of life on Earth. Numbers of species, genera, families, and higher orders in the biological classification scheme all follow the same upward trend. It would take an entire book, or more, to map out what is known about the details of that evolution, and in the remainder of this chapter I will touch on only a few of the highlights. We will work our way from the time of the first tetrapods toward the present by examining some of the details the rock record reveals about the evolution of both life and the planet itself. You may want to refer to the time line in figure 31 as we do this, especially if you are not very familiar with the geological timescale.

The Carboniferous period, when the early tetrapods flourished, takes its name from the abundance of carbon contained in its rocks. Much of this carbon is in the form of coal. There are bits and pieces of coal in some older sedimentary rocks, but vast quantities of it occur in Carboniferous rocks in Europe and North America. Why the sudden appearance of so much coal? And why in those places in particular? Answers to those questions follow neatly from the story of plant evolution and the workings of plate tectonics.

During the Carboniferous, North America and much of what is now Europe were joined together in a large continent that straddled the equator. Early plants had by then diversified and developed thick, sturdy trunks that allowed them to grow into trees tens of feet high. Especially along coastlines in tropical areas, there were huge "coal swamps," populated by forests of tall trees, giant ferns, and many other types of plants. These prolific environments produced prodigious amounts of organic debris, and in many places it simply piled up into thick deposits of peat. Buried, compressed, and heated over millions of years, it became coal.

But the Carboniferous plants didn't just grow in coal swamps; they were also busily diversifying and occupying many new environments.

For the first time in its history, the Earth developed true soils of the type we know today, full of decaying plant material, bacteria, and small crawling creatures. Plant roots and organic acids helped break down the solid rock of the land surface, accelerating soil formation. The rapid proliferation and growth of plants also had another effect on the planet: through photosynthesis, green plants took carbon dioxide from the atmosphere and transformed it into organic carbon. Much of this carbon was more or less permanently removed from the atmosphere and stored as coal or in soils, and the carbon dioxide content of the atmosphere decreased.

The coal deposits of the Carboniferous have an interesting feature: they usually occur as alternating layers of coal and marine sedimentary rocks. Because of their economic importance, these deposits have been studied in considerable detail, and the origin of the alternating layers is quite well understood—they were produced by fluctuating sea levels. When sea level was low, coastal swamps thrived; when sea level rose, the swamps were flooded and the organic debris was buried under ocean sediments. And the reason sea level rose and fell? Almost certainly it was due to cycles of glaciation much like those that have caused sea level to fluctuate during the Pleistocene Ice Age. As the proliferating Carboniferous plants took carbon dioxide out of the atmosphere, the greenhouse effect weakened and temperatures dropped. Gondwana was still near the South Pole, and permanent ice began to form there. Increased albedo caused further cooling, and the Milankovitch astronomical cycles described in chapter 8 triggered alternate advances and retreats of the ice, resulting in the sea level changes that formed the interlayered coal deposits.

The Carboniferous ice age was at least as extensive and certainly longer-lived (it continued into the Permian period) than the ice age at the end of the Ordovician. Glacial deposits from the Carboniferous-Permian episode occur in southern Africa, South America, Australia, and India. They were part of the evidence used by Alfred Wegener when he developed his theory of continental drift. But in spite of its duration and geographical reach, the Carboniferous-Permian ice age does not coincide with a mass extinction, perhaps because its onset was gradual and the initial sea level decline was not as disastrous (in terms of habitat destruction) as that at the end of the Ordovician.

At the end of the Permian period, however, came the largest of the big five extinctions, in which most life on Earth was wiped out. As outlined in chapter 10, most of the available evidence suggests that the extinctions took place in stages, in response to relatively gradual environmental changes followed by one or more short-lived catastrophic events that were probably linked to the eruption of the Siberian flood basalt province.

The supercontinent Pangea was fully formed when the P-T extinction occurred. Pangea was long and narrow, with Gondwana, comprising today's southern continents, in the south, and North America, Europe, and much of Asia to the north (see figure 33), making it possible, it has been said, to travel from pole to pole on land. But in the immediate aftermath of the extinction, there were few animals left to wander the vast supercontinent, nor were there many populating the large, single ocean that surrounded it. The crisis was so severe that recovery to previous levels of biological diversity was slow—much slower than after other large Phanerozoic extinctions. One indication of its severity is the so-called Lazarus species or genera, named after the biblical figure who is said to have returned from the dead. The Lazarus fossils are of animals that completely disappeared from the fossil record at the P-T boundary, only to reappear again after a long time gap. The most likely explanation for this is that the Lazarus organisms barely survived the extinction, clinging on only in geographically isolated patches and with greatly reduced numbers, and that it took a long time for them to again become abundant enough to be preserved as fossils.

The P-T extinction killed off many of the amphibians and reptiles that had evolved from the first tetrapods. But some twenty million years after the extinction, by about 230 million years ago, the dinosaurs had appeared. A fossil of that age from South America—of a small,

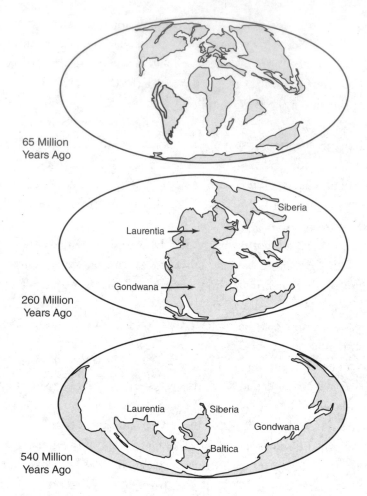

65 Million
Years Ago

260 Million
Years Ago

540 Million
Years Ago

Siberia

Laurentia →

Gondwana →

Laurentia   Siberia

Gondwana

Baltica

Figure 33. The positions of the continents at the beginning of
the Cambrian period (bottom), close to the Permian-Triassic
boundary (middle), and at the K-T boundary (top). In the earliest map, Laurentia contains much of present-day North America; Baltica contains Scandinavia, eastern Europe, and parts of
western Russia; and Gondwana contains the current southern
continents (Africa, Australia, Antarctica, and South America)
and the Indian subcontinent. By the time of the K-T boundary, continent shapes are quite recognizable. (After maps by Ron
Blakey, Northern Arizona University; see http://jan.ucc.nau
.edu/~rcb7/.)

meat-eating creature about three feet long and weighing around twenty pounds—appears to be a dinosaur, or an immediate predecessor. The transition from reptiles to "mammal-like reptiles" and then to true mammals was occurring at about the same time. Paleontologists argue about exactly what characteristics define a true mammal and when these characteristics first appeared, but by late in the Triassic period (between about 210 and 205 million years ago) small animals had evolved that were either mammals or their immediate ancestors. One of these was found in 1991, beautifully preserved, in rocks from Yunnan, China, that are about 195 million years old. The creature was tiny, weighing less than a tenth of an ounce (for comparison, that is roughly the weight of the smallest living bat), and it came from a mammal lineage that is now extinct. This small animal would not have been much competition for dinosaurs, which diversified rapidly and became the dominant vertebrates for more than 150 million years. But when the dinosaurs became extinct at the K-T boundary, mammals evolved rapidly, getting much larger and expanding into new habitats.

Not long after the P-T extinction, Pangea began to break up. Over the following 225 million years or so, continuing up to the present, the continents we know today broke off, drifted across the globe, and sometimes collided to push up mountain ranges, including the Himalayas and the Alps. The breakup began with incipient rifts in the crust of Pangea, probably caused by hot upwelling material in the mantle that bulged up the overlying lithosphere, stretching it until it cracked. One of these, a several-pronged rift, broke apart Africa, North America, and South America, and eventually evolved into the present-day Atlantic Ocean. Initially the rifts were similar to the great rift valley of Africa today, steep-walled, roughly linear valleys within the continent, characterized by volcanic activity. However, as the process continued, the rifts widened and connected to the surrounding ocean. Seawater poured in to fill the valleys. But this was a sporadic process—the rifts were narrow enough that lava flows or uplift of the shoreline sometimes blocked the connection with the ocean, and when this happened the seawater completely evaporated away, leaving behind only a deposit of salt. Then the ocean would again flood in, and the process was repeated. These episodes are recorded in the thick submarine salt deposits along the coasts of Africa and South America and in the Gulf of Mexico, all of them formed early in the breakup of Pangea.

Rifting of the northern parts of Pangea occurred much later; Europe and Greenland, for example, only started to split apart about sixty million years ago, creating the Norwegian Sea. In the Southern Hemisphere, Gondwana had also begun to split apart. Antarctica remained near the South Pole, a position it has occupied for more than 200 million years. However, about 130 million years ago India, still attached to Madagascar, broke off and began to travel rapidly northward, and about fifty million years ago it approached and began to collide with Asia, as described in chapter 5. Australia did not begin to separate from Antarctica until about eighty million years ago, and at first its northward travel was slow. But for the past forty million years or so, it has been moving more rapidly, and its current trajectory suggests that it will collide with Southeast Asia at some point in the future, suturing new continental fragments onto its northern margin.

In yet another example of the interconnectedness of processes on Earth, the breakup of Pangea had a direct impact on evolution. As biological recovery began after the massive Permian-Triassic extinction, there was, effectively, still only one continent and one ocean. In principle, marine creatures could populate a continuous coastline without any significant geographical barriers. On land, there were obstacles such as mountain ranges and large rivers, but no insurmountable physical barriers preventing animals from colonizing most of the continent. All that changed as Pangea began to split up.

The mammals present a striking example. Before Pangea broke apart, marsupial mammals—those with a short period of gestation followed by prolonged residence in the mother's protective pouch—were abundant in the northern parts of Pangea (present-day North America, Asia, and Europe). The oldest marsupial fossil yet found, a beautifully preserved skeleton of a small, tree-climbing creature that retains delicate impressions of its furry coat, is about 125 million years old. It was found in China, and many paleontologists believe that the divergence between placental and marsupial mammals occurred in Asia early in the Cretaceous period, not long before this fossil marsupial lived. Regardless of exactly where and when marsupials originated, however, they migrated to parts of the southern continent of Gondwana before it began to break up.

Today, largely because of the way in which the sequential fragmentation of Pangea occurred, marsupials are restricted almost exclusively to Australia and South America. Australia is a particularly clear example. Marsupials had established a foothold on the continent before it broke off from Antarctica, and once separation was complete, Australia was geographically isolated. For tens of millions of years, no new mammal species could join the indigenous marsupials, and they evolved and specialized to fill the same types of ecological niches occupied by placental mammals on other continents. As Australia's plate tectonic journey brought it closer to Indonesia in the relatively recent past, a few small placental mammals from Asia (bats and small rodents, for example) managed to make the relatively short water crossing, but marsupials continue to dominate the mammalian fauna of the continent to this day. Because of their long isolation, most other Australian animal groups—reptiles, amphibians, insects, and even fish—are also unique to the continent.

There is one environment I have not really discussed up to this point: the air. Insects—as we saw earlier—were buzzing around off the ground at least 400 million years ago, during the Devonian period. But larger animals didn't take to the skies until much later. The first were the pterosaurs, sometimes called "flying lizards," which appear in the fossil record about 220 million years ago, during the Triassic period. Pterosaurs flew the Earth's skies for more than 150 million years, and by late in the Cretaceous period some of these reptiles had become gigantic—fossilized individuals have been found with wingspans of thirty feet.

However, the pterosaurs met their demise in the K-T extinction. So where, then, did the current denizens of the air, the birds, come from? This is a problem that has intrigued paleontologists for many years. From the fossil trail, it seems clear that the ancestors of the earliest birds were small, agile dinosaurs called "theropods." Although they couldn't fly, these dinosaurs ran on their hind legs, and some developed relatively long "arms" and hands. They also had the same light, hollow bones that characterize birds, which is one of the reasons they were so agile. Analysis of their skeletons suggests they had a muscle structure that enabled them to bring their arms and hands together to grab a flying insect, in a motion akin to the stroke of a bird's wing. Some theropods also developed feathers, perhaps for insulation, and having feathers on their arms might have helped them leap or glide in pursuit of prey. Some small dinosaurs may also have glided from trees, much like flying squirrels. Such ideas are still speculative, but they are based on real evidence provided by fossils.

The oldest known true bird lived 150 million years ago, near the end of the Jurassic period, when dinosaurs were still very abundant. Its fossilized remains were first discovered in a limestone quarry in southern Germany in 1861, and although additional fossils have been recovered since then, all from the same limestone formation, no earlier bird fossil has been found. The creature, named *Archaeopteryx*, had broad wings and a long tail, and the feathers on its wings were asymmetric, a crucial characteristic of feathers developed for flight. This early bird was fairly small; the largest individuals were probably not much more than a foot and a half long. The extent of its flying skills is not known—it could certainly take to the air, but it did not have the flying ability of most modern birds. Whether it was primarily a glider, or whether it could also propel itself along with crude wing flaps, is uncertain.

Archaeopteryx played an important role in early debates about evolution. Thomas Huxley, a nineteenth-century British biologist, was the first to link the origin of birds to dinosaurs, and he saw *Archaeopteryx* as a "missing link." He also claimed that the fossil provided compelling evidence that Darwin—who had published *Origin of Species* just a few years before *Archaeopteryx* was discovered—was right (Huxley is sometimes referred to as "Darwin's bulldog" because of his advocacy for Darwin's work). Since the 1990s many new fossils have been discovered that fill in some of the gaps between *Archaeopteryx* and both earlier theropod dinosaurs and later, more evolved birds. Huxley's evolutionary link is no longer in question.

Once dinosaurs learned to fly and became birds, they diversified rapidly and adapted to a wide range of food sources and environments. Some, in isolated environments with few predators, even lost their ability to fly, like today's penguins. Others became prodigious fliers. In today's world, the record goes to the sooty shearwater, a seabird that each year completes a long, looping migration between the Northern and Southern Hemispheres in a round trip estimated to be an incredible forty thousand miles. The shearwater—and some other seabirds—can also "fly" rapidly underwater in pursuit of prey.

Somewhat paradoxically, the last great extinction of the Phanerozoic, at the K-T boundary, wiped out the dinosaurs but not their avian descendents. Like all plant and animal groups, birds were affected—several groups that had successfully diversified during the Cretaceous period became extinct—but to a lesser extent than most other life forms. Their high survival rate is still a puzzle to paleontologists, especially when even small dinosaurs, which were probably about the same size as some of the larger surviving birds and occupied similar environments, didn't make it. Whatever the answer is to that puzzle, the demise of the dinosaurs opened up the field for birds to expand, as it did for mammals. Recovery was not immediate for any surviving groups, plant or animal, but within a few million years the stage was set for the evolutionary march toward today's breathtaking array of living things.

Sixty-five million years, the length of the last era of the Phanerozoic, the Cenozoic, is not a particularly long stretch of time in geological terms, but even this interval has brought many changes to our planet. From the decimated biological world in the aftermath of the K-T extinction arose modern birds, ranging from the inconspicuous wren to the brilliantly colored toucan with its giant beak, and from the gangly ostrich to the darting hummingbird. Mammals too have reached astounding diversity, from huge whales to bats and tiny shrews—and humans. The Himalayas and the Alps were pushed up as drifting pieces of Pangea collided, and thirty-five million years ago the climate began to cool steadily toward the Pleistocene Ice Age of the past few million years. Very recently we entered what Nobel Prize winner Paul Crutzen termed the "Anthropocene," that part of the Earth's history when humans became a significant geological force, with the capacity to change (wittingly or not) the course of our planet's history. Climate change and the "sixth extinction" are present-day manifestations of that capacity. But the changes need not all be negative. Humans also have the capability to predict, and to protect. To a considerable extent, the immediate future of our planet is now in our own hands.

CHAPTER THIRTEEN

Why Geology Matters

The preceding twelve chapters have, I hope, provided a window into the multifaceted world of earth science. The field got its start long before the word *geology* was even coined, as a purely practical endeavor to locate and extract materials from the Earth for human use: chert and obsidian for cutting blades and weapons, hematite (iron ore) for early "crayons" and body paint pigment, clay for pottery, iron and copper for tools and ornaments. Today almost everything we use in our daily lives comes from the Earth, and geology plays a crucial role in feeding our voracious appetite for materials and energy. But the field of earth science has grown far beyond its practical beginnings. In the broadest sense, it is an intellectual pursuit that encompasses studies of the atmosphere, the oceans, the solid earth, and even other planets of the solar system. It is perhaps the most truly interdisciplinary of all the sciences, drawing on the efforts of mathematicians, physicists, chemists, engineers, and biologists. All of these scientists are ultimately pursuing the same goal: a better understanding of the Earth's natural processes. They may work in an obscure corner of the field or on a grand synthesis like plate tectonics, but they are continually revealing fresh details about how our planet operates. In this pursuit, new subdisciplines keep popping up—for example, the emerging area of "biogeochemistry," which, as its name implies, focuses on the intersections between chemistry, biology, and geology. The reason earth science matters so much today is that the field holds the keys to understanding and addressing many of the most pressing problems facing society.

What are those challenges? The issues are many and diverse, but among the most important are: availability of energy and mineral resources; access to fresh water; climate change; ocean acidification; agricultural sustainability; and maintaining biodiversity. Many of these topics are interconnected, and some of them have been touched on earlier in this book. You might argue that the last two, especially, are not simply "earth science" issues, and you would be partly right. But they too require input from the geosciences. Paleontologists, for example, from their knowledge of evolution and extinction, have a unique perspective on present-day biodiversity. Many problems in agriculture involve an understanding of soil science and erosional processes—both integral parts of the earth sciences. As I have tried to emphasize throughout, a holistic view of our planet is important for fully understanding the workings of the Earth today, for decoding its history, and also for using that knowledge to predict the future. This is not a new idea, but it is one that has gained widespread acceptance only over the past few decades. It is increasingly clear that there are many links among Earth processes that may not be intuitively obvious; for example, oceanographers today need to know something about atmospheric chemistry, mineralogy, and the weathering of rocks on land to understand ocean chemistry, and paleontologists need to know about plate tectonics, atmospheric chemistry, and how ocean chemistry has changed through time in order to understand the fossil record. Different parts of the Earth can no longer be considered in isolation.

Mineralogy illustrates this concept well. As a discipline, it is often considered dry. Generations of geology students have been required to learn the properties of various minerals in order to identify them, and sometimes they wonder about the point of it all. When I was a student, our department had a famous (or perhaps infamous) exam: the white mineral test. All the samples were white, and many were rare white varieties of minerals that usually occur in a different color; we had to identify the minerals on the basis of properties other than color. It was not easy. The exam underscored that one has to consider all aspects of a sample in order to identify it, but we wondered if it was really a useful exercise. In 2008, however, Robert Hazen of the Carnegie Institution for Science in Washington, D.C., and his colleagues made a startling proposal that sparked widespread interest in mineralogy and may ensure that future generations of students will view the field in a different light; it also put minerals at the heart of the concept of the Earth as an integrated whole. Minerals, Hazen and his colleagues said, have evolved along with the Earth itself.

There is some debate about whether "evolution" is the right concept to use for minerals. But Hazen and his co-workers pointed out that there were only about sixty different minerals in the materials that accreted to form the asteroids and the planets (this is known from studies of meteorites). On the early Earth, through processes such as melting, volcanism, metamorphism, chemical segregation in the planet's interior, and the formation of both oceanic and continental crust, the mineral total rose to about 1,500. Today the tally is approximately 4,300, largely, these authors conclude, because of the evolution of life. Mineral diversity has increased as life and the planet have evolved together over geological time. This is a different and compelling take on mineralogy.

That life and mineral diversity are so closely linked might at first seem surprising. There are ninety-two different elements in the periodic table, and in principle they could combine to form an almost infinite number of possible compounds. But the compounds we know as minerals only form spontaneously under very specific conditions; especially important are the temperature, the pressure, and the concentrations of various chemical elements in the surroundings. On the early Earth, the material that accreted to our planet was subjected to new, very different, and varying conditions, interacting with atmospheric gases and water at the surface and experiencing a wide range of different temperatures and pressures in the interior. As a result, the number of mineral species increased rapidly.

When life appeared on Earth, a new and potent mechanism became possible: biologically mediated mineral formation. Organisms make minerals that are useful (teeth, for example, are made of the mineral apatite, which is the compound calcium phosphate) but would not form spontaneously. The earliest known biologically formed minerals occur in stromatolites (see figure 10). These massed colonies of microbes trap preexisting mineral grains as they build up their structures, but they also precipitate several minerals to make the intricate, fine-scale layering that characterizes them. In addition to forming new minerals, life also affected the mineral tally indirectly: once photosynthesizing plants gained a foothold and began to oxygenate the atmosphere and oceans, a whole new array of oxide minerals appeared at the Earth's surface. The increase in mineral numbers is irreversible; we can never go back to the sixty or so that existed before planet formation.

Research in mineralogy has both a practical side (did you know that the development of "clumping" cat litter depended on mineralogical expertise?) and a "pure research" aspect, and this is true of most areas of earth science. Although many geoscientists (and the agencies that support their research) focus on practical earth-science-related issues facing society—issues that are clearly of utmost importance and deserve all the expertise that geoscientists can offer—the purely curiosity-driven search for knowledge about the Earth, fortunately, continues to flourish. Much of the material in this book has dealt with this kind of research—the kind that has produced truly revolutionary ideas about our planet, like plate tectonics or evolution (I count Darwin as an earth scientist here; he was fascinated by the subject, and fossils and geology were important parts of his thinking). But here I would like to examine some of the more applied areas of earth science research, specifically those related to present and future societal concerns.

To some extent, *applied* and *pure* are somewhat arbitrary, although useful, terms for describing research, because the two are frequently closely intertwined. Take, for example, nuclear weapons, a topic you would probably not normally associate with earth science, except perhaps in the sense that finding and mining uranium requires the skills of geologists and geological engineers. But a crucial part of nuclear weapons programs is testing. And because most countries want to reduce and eventually completely eliminate the threat of nuclear weapons, and are at the same time often suspicious about the intentions of other nations, finding ways to detect and accurately characterize secret nuclear tests has become an important priority. To a large degree the task has fallen to the earth science community, especially seismologists, but also geochemists, oceanographers, and atmospheric scientists.

International negotiations aimed at banning nuclear weapons testing began during the Cold War, and geoscientists have been involved almost from the start. Neither of the two original main protagonists in these negotiations, the United States and the Soviet Union, wanted teams of scientists wandering around their countries inspecting their classified nuclear facilities. But while Soviet scientists claimed that they could easily identify even small explosions at large distances, their American counterparts insisted that existing seismic capabilities were insufficient to detect even quite large nuclear explosions unless monitoring seismographs could be placed close to test sites. The conflicting claims raised suspicions on both sides that politics was trumping science. In the United States, the result was the creation of a large government-funded program designed to improve seismic detection of nuclear explosions. The effort, initiated in the 1960s, was highly successful, in more ways than one. Because seismologists were given free rein in their research, the program not only greatly advanced nuclear test detection, it also played an important role in the development of plate tectonics theory and improved understanding of earthquakes and the interior structure of the Earth. Interestingly, it turned out that the original discrepancy between Soviet and U.S. detection capabilities was real and the science had been accurately reported by both sides: the disparity resulted because seismic waves propagate differently in the different geological settings at U.S. and Soviet test sites.

Although seismology is at the heart of detection methods, there are also other earth-science-based monitoring methods now in use. Hydrophones and barographs, instruments designed to record pressure waves in the ocean and atmosphere, respectively, can pick up signals from distant explosions, whether they originate from undersea volcanoes, meteorites entering the atmosphere, or nuclear weapons tests. Also, because nuclear explosions release characteristic radioactive isotopes into the environment (this happens even when tests are carried out underground), detection of even very small amounts of these isotopes provides unambiguous proof that a nuclear explosion has occurred. Thus detection capabilities developed within the earth science community, mainly for other purposes, have been vital for moving forward the Comprehensive Nuclear Test Ban Treaty, which was adopted by the United Nations in 1996. When it is eventually fully implemented, the treaty will ban all nuclear testing, and the same earth science expertise—coordinated through a United Nations organization headquartered in Vienna—will insure that all countries adhere to the ban. United States President Barack Obama has vowed to work toward a nuclear-weapon-free world. Thanks in large part to the geoscience community, the ability to verify when that goal is attained now exists.

Hopefully, we will one day be able to scratch nuclear weapons from the list of threats to the future of humankind. But there are still other issues that are of concern—even if several of the potentially civilization-destroying natural hazards discussed earlier in this book, such as supervolcanoes and large asteroid impacts, are unlikely to affect the Earth in the near future. In what follows, I would like to address a few of these issues from the list at the beginning of this chapter.

Let's begin with energy and mineral resources. Broadly speaking, the solid Earth supplies most of the things we need for modern society to function, in the form of minerals or fossil fuels. It is a cliché to say that these resources are limited, but it is also true. The case of oil and gas has been much discussed in the media, and even though the time of "peak oil," when production reaches its maximum and begins an irreversible decline, keeps getting pushed into the future because of new discoveries and better recovery technology, it will eventually arrive. There is a reasonable chance that other, renewable energy sources will be developed to meet society's needs before that happens. But there is also a less widely discussed problem: diminishing mineral resources.

All of the minerals we use come from the Earth's crust, which is the only part of our planet that is readily accessible (and most of them come only from the continental part of the crust). However, just eight of the ninety-two elements in the periodic table account for 99 percent of the weight of the crust. Among these eight elements are aluminum and iron, both obviously important for modern society, but most other crucial elements, from antimony to zinc, make up only tiny fractions of the total. Still, even these tiny fractions add up to impressive absolute amounts. In the nineteenth and early twentieth centuries, enterprising con men used that knowledge to persuade people they could get rich by extracting gold from seawater. There *is* gold dissolved in the oceans; modern analyses show that the total amount is somewhere between three million and thirty million tons, equivalent to more than a thousand years' worth of gold production from mines at current levels. But the problem is that the concentration—the key parameter that determines whether a resource is economically recoverable—is extremely low. A ton of seawater contains just a little more than a billionth of an ounce of gold. You would have to process a very large amount of water to get rich. Needless to say, those who invested in seawater schemes never saw their money again.

On average, the continental crust contains much higher concentrations of gold than seawater does, but still far too little to make recovery feasible from just any old rock—otherwise gold mines would dot the landscape. And what is true for gold is true for all mineral resources: there must be a concentration mechanism to make mining viable. Fortunately for us, there are many processes that have acted through geological time to separate and segregate various chemical elements and minerals into deposits that can be mined economically. As far as we know, these processes are unique to the Earth; the geological makeup and history of the Moon and the terrestrial planets make it unlikely that Earth-like deposits will ever be found on these objects.

What kinds of processes sweep up rare chemical elements from large volumes of the Earth and concentrate them in ore deposits? They are numerous and varied, but important in many of them is the presence of water.

Water is a potent solvent; it can carry many different chemical elements and compounds in dissolved form. In most cases, the higher the pressure and temperature, the more dissolved material it can hold. When water circulates through hot parts of the Earth's crust, it leaches out large quantities of various elements from the surrounding rocks and transports them over long distances. The ability of these so-called hydrothermal solutions to carry dissolved metals is enhanced by the presence of other elements that form chemical complexes with the metals, such as chlorine and sulfur. Complex solutions rich in these components emerge today from hot springs along the ocean ridges. Known as "black smokers," these springs occur where plates are splitting apart and magma wells up to create new ocean crust. Seawater seeps into the rocks, is heated to very high temperatures as it circulates, and eventually spews back out onto the seafloor in a hot, buoyant jet. At some black smokers, water temperatures of well over six hundred degrees Fahrenheit have been measured (the crushing pressure of the deep ocean prevents the water from boiling even at such elevated temperatures). But as soon as these superheated, crystal-clear hydrothermal solutions mix with cold seawater, they lose their ability to hold dissolved metals. Small, black particles of metal sulfide nucleate spontaneously, giving the hot springs the appearance of factory chimneys belching black smoke. The tiny grains that precipitate are rich in copper, zinc, lead, nickel, iron, and several other metals, and they blanket the seafloor with a high-grade ore deposit.

The black smokers were discovered serendipitously in 1979 by geologists using the research submersible *Alvin* to investigate the ocean ridge system. Since then they have been found in many other places, and preliminary attempts have been made to mine some of them by vacuuming up the fine, loose sediment. However, at-sea mining operations are messy and very expensive and can be environmentally damaging, and in many places the black smokers are too dispersed to form deposits on the scale required to make mining economical. But in an echo of the early geologists' maxim that "the present is the key to the past," discovery and observation of this present-day ore-forming process has provided insight into how some types of ancient, on-land mineral deposits originally formed—and has made exploration for them easier.

Not all mineral deposits are formed by black smokers, of course. Many different processes can concentrate valuable minerals: gravity segregates layers of heavy, economically important minerals at the bottom of pools of magma injected into the continental crust; rainwater leaches away easily soluble minerals, leaving behind insoluble but valuable ones; heat in volcanically active regions—especially above subduction zones—provides the means for distilling out various valuable elements that are deposited in zones or veins in the surrounding, cooler rocks.

Ore bodies created by all of these processes were first discovered by accident thousands of years ago, but observant people quickly realized that there are recognizable associations and that certain ores always occur together with specific rock types. Such observations have made it easier to prospect for new deposits. In some cases, however, the Earth's most obvious mineral deposits are already "mined out." Finding new ones is increasingly difficult, and many nations have established geological surveys to map and assess their country's mineral and energy resources. In resource-rich countries such as the United States, Russia, Canada, Australia, and several others, these organizations have large research arms, also with a significant focus on resources. But in the twenty-first century, encouraging exploration and aiding the mineral industry is a double-edged sword. On the one hand, we need a supply of rare elements for the manufacture of everything from airplane engines to lightbulbs. Modern society simply could not function without them. On the other hand, there is no question that the mining industry has sometimes been guilty of exploiting local populations and damaging the environment in pursuit of profit. Unchecked, such problems could intensify in the future, because mineral resources are indisputably finite.

Is there a solution to this dilemma? Is it possible to have a sustainable future in which society's needs are met without irreparable damage to the Earth and its people? The jury is still out, but there are some hopeful signs. One is that in spite of rapidly increasing consumption, the world's "reserves" of many minerals have increased dramatically over the past fifty years. How can this be? Reserves are simply the existing, known quantities of particular minerals. They have increased because advances in exploration technology have made prospecting for mineral deposits more efficient, and also because research has increased our understanding of how the deposits form, and has guided exploration for them to the most likely geological settings. Reserves do not include estimates of undiscovered deposits, which will continue to be found— although probably with diminishing frequency. Formal estimates of reserves also do not take into account the possibility that technological advances will permit economical extraction of minerals from rocks that were previously deemed to hold too little to be useful. This has occurred repeatedly in the history of mining—even the "tailings," the residue from prior mining operations, have often proved to be a rich source with the advent of new extraction methods. Computer modeling and visualization have also played a role in increasing known reserves by enabling geologists to create virtual extensions of existing mines and project the locations of ore zones far beyond existing mine excavations.

At some point, however, few or no new deposits of specific valuable minerals will remain to be discovered. Is it possible to sustain adequate reserves over long time periods in such a situation? One strategy is to substitute a relatively abundant element for a rarer one. The use of aluminum in place of copper for electrical transmission is a good example. And for many rare metals, recycling is likely to play a pivotal role in maintaining reserves in the future. Scrap steel, copper wire, and aluminum cans have long been recycled because they are relatively pure and can be processed easily. For more complex materials like electronics, the infrastructure for reclaiming metals is not yet highly developed. But as competition for resources increases, that is likely to change. A few years ago the U.S. Geological Survey (USGS) examined the potential for recovering metals from recycled cell phones. These everyday objects contain gold, silver, platinum, and other valuable metals, and could be an "ore" of the future.

As this is written, at least four billion mobile phones are in use around the world, and the number is increasing rapidly. The USGS study concentrated on the United States, where phones are replaced, on average, every year and a half. The study was based on data from 2005, and it concluded that metals in discarded phones that year had a value of $82 million. While cell phones have gotten smaller since then, their metal contents have not changed drastically. And both gold and silver have more than doubled in price since 2005. Prices for other metals in the phones have also risen substantially, greatly increasing the recycling value. Some of the data in the USGS report came from a nickel-mining company, Falconbridge, which in 2005 had already set up an operation to recover metals from recycled cell phones.

It will never be possible to attain anything like 100-percent efficiency in mineral recycling, so it is possible that some nonrenewable resources will eventually become exceedingly scarce. But it may be possible to push that day of reckoning very far into the future. Even previously exploitative mining companies seem to recognize that it is in their best interest to operate efficiently, minimize environmental degradation, and even remediate disused, contaminated operating sites. For many years, for example, sulfur-rich gas from a smelting plant run by Falconbridge killed vegetation and damaged health over a wide area downwind from the plant. When the company finally did something about it by capturing the sulfur, they were able to turn it into sulfuric acid, which they could sell at a profit. Not every technological fix will be so easy, or so profitable. However, it does seem that with the right approaches, using current and developing technology, it may be possible to meet society's needs for mineral resources with an acceptable level of environmental disruption for an extended period in the future.

Access to water is another geoscience-related issue that is increasingly in the spotlight. Some earth scientists who once focused their attention on mineral and energy deposits are now turning their expertise to this valuable resource; a few of them call it the "new gold." Water for irrigation, hydroelectric power generation, industry, and everyday household use has long been an issue in arid regions, and in some places it is fast becoming a national security problem. In Iraq, for example, all water comes from rivers that flow into the country from elsewhere—and are therefore ultimately controlled by other nations. As population growth and economic prosperity increase the demand for water, and as climate change threatens to disrupt historical patterns of precipitation and evaporation, water availability and quality is fast becoming a worldwide problem.

Nearly all of the world's water resides in the oceans, but for most of us—surfers and sailors and fishermen aside—what is important is fresh water. That fresh water, however, ultimately comes from the oceans; about 100,000 cubic miles of seawater evaporates every year, and it all falls, somewhere, as fresh water: as snow on the glaciers of Greenland or the ski resorts of the Rockies, as rain that waters our lawns and replenishes our rivers and lakes, or as precipitation that falls right back into the ocean. In geological terms, the water cycle operates on a short timescale. On average, a molecule of water spends only about three thousand years in the ocean before it is evaporated into the atmosphere, and about a week later it finds itself part of a raindrop falling back to the Earth's surface.

At this instant in geological history, most (about two-thirds) of the

Earth's fresh water is stored as ice in glaciers, nearly all of it in the Antarctic. As we have seen in other parts of this book, the volume of glacial ice has fluctuated widely through geological time, with important consequences for sea level. But if present-day glaciers hold just two-thirds of the world's fresh water, where is the rest? You might think, reasonably enough, that it resides in lakes and rivers, but you would be wrong. About 99 percent of it is out of sight, below ground. Underground water is a precious resource, because it is usually very pure, and also because it is often present even in areas where the surface is arid. It is, in principle, a renewable resource. But it is only sustainable if it is managed so that water withdrawal does not exceed the amount added to the system.

In the United States, the High Plains region in the central part of the country is underlain by a vast reservoir of fresh water that has made it one of the country's prime agricultural areas. The topography, soil, and climate of the High Plains are excellent for farming and grazing, but rainfall is sparse, and the lack of precipitation was a serious impediment to agriculture in the past. Although the first wells were drilled into the underlying aquifer in the early part of the twentieth century, large-scale pumping of water for irrigation did not take hold until the 1940s and 1950s. Initially, there seemed to be an inexhaustible supply, but as more and more water was taken out, the water table dropped, wells had to be deepened, and in a few places it is no longer possible to extract useful quantities for irrigation. Clearly, replenishment has not kept pace with water withdrawal.

The USGS has identified sixty-two "principal" groundwater reservoirs in the country as important national resources; the High Plains aquifer is one of them. It is tempting to think of aquifers as underground rivers, but the reality is very different. Some are solid rock formations; others are made up of unconsolidated sand and gravel that have not yet been compacted into solid rock. All are permeable—they have interconnected pore spaces that water can flow through, in much the same way it flows through a sponge, slowly and following convoluted paths.

In the absence of human intervention, water seeps into the ground at high elevations (the recharge area), flows through an aquifer, and exits in springs and seeps at lower elevations. Gravity is the force that keeps the water moving.

The High Plains aquifer is largely made up of unconsolidated sand and gravel eroded from the Rockies millions of years ago and deposited to the east, where it filled up river-cut valleys and spread out as gently sloping layers that are now deeply buried. The USGS studies have found that the "age" of water in the aquifer—essentially the length of time it has spent underground—is more than ten thousand years in the deepest parts of the system. That ancient water originated in melting Rocky Mountain glaciers at the end of the last glacial period of the Pleistocene Ice Age, and it has obviously traveled very slowly through the aquifer. Its age indicates that recharge rates, at least for that part of the aquifer, are very low.

A crucial task for geoscientists managing groundwater resources is measuring the recharge rates accurately so that sustainable withdrawal rates can be set. This is not an easy undertaking, because the regions that feed aquifers are often large, encompass different types of topography, and may experience varying levels of precipitation. At the simplest level, monitoring well levels provides a rough measure of the balance between water entering and leaving an aquifer, but more quantitative approaches involving computer models of entire aquifer systems are now the norm, as are strategies to store "excess" surface water in natural aquifers rather than allowing it to disappear through evaporation or runoff. Most of us take our morning shower for granted, but in many places that is only possible because geoscientists and engineers have worked out the details of buried aquifers and their recharge areas, and have developed effective plans to make the flow of water from our taps sustainable.

Those strategies may have to be modified because of climate change, however. As the Earth warms, the water cycle will intensify: warmer temperatures increase the rate of evaporation, which will in turn feed more extreme precipitation. In effect, the whole cycle will speed up, with more water being processed through the atmosphere. This is likely to increase the frequency and severity of both storms and drought, and will also alter patterns of precipitation around the globe in ways that are difficult to predict.

Global warming will also affect the water cycle in a way that may not be quite so obvious but will have an equal or greater impact on humanity. In many parts of the world, glaciers are the primary source for fresh water, either as groundwater or as surface runoff. Glaciers in the Himalayas and the Andes are especially important; they provide water for hundreds of millions of people. But these great storehouses of fresh water are shrinking rapidly, and many of them will eventually disappear entirely as the climate warms.

In the short term, the rapid melting of glaciers feeds abundant fresh water into below ground aquifers and surface streams, giving a false sense of their long-term potential not only for water supply, but also for projects such as hydroelectric power generation. In the longer term, though—and in the case of some Himalayan glaciers, the longer term will be measured in decades to at most a few hundred years—the disappearing glaciers have serious implications. The water supplies for large segments of the population in India, Pakistan, China, and other parts of Southeast Asia hang in the balance. The situation also raises important questions yet to be resolved by international law. Who "owns" the water of a glacier that is located in Tibet but feeds rivers that flow into India and, eventually, Bangladesh? And what about aquifers that cross national borders? Such issues are difficult to settle even among the most amicable of neighbors. In a region where there is a long history of animosity, they could be inflammatory. Regardless of legal questions and national boundaries, however, an urgent task for geoscientists is to understand the regional details of the water cycle, and how they will change as the climate warms, in order to make access to adequate water resources sustainable for the world's growing population.

Finally, let's turn to the geoscience issue that seems to be on everybody's mind: climate change. Without question, this is one of the central issues—perhaps *the* central issue—for modern earth scientists, and indeed for everyone on the planet. A closely related phenomenon that has received much less attention is ocean acidification, which, as we saw briefly in the discussion of the Great Warming of 55 million years ago (in chapter 9), is also caused by an increase in atmospheric carbon dioxide.

As an earth scientist, I am often asked if I "believe" in global warming, or whether it is really true that human activity is changing the Earth's climate. Sometimes this leads to very long—even exhausting—discussions. I have no problem with skeptics; questioning assumptions is a good way for both parties in a discussion to learn. But I do take issue with those who either twist facts to fit an agenda, or simply refuse to acknowledge the truth of well-documented observations. When it comes to topics like climate change or evolution, there seems to be a large number of people in this category.

In the case of human influence on climate, there are a few indisputable facts. The first is that carbon dioxide in the atmosphere traps heat that would otherwise radiate out into space. That is a principle of physical chemistry, and it was first described by John Tyndall, a British scientist, who measured the effect in the mid-1800s. The second indisputable fact is that human activity has drastically increased the amount of carbon dioxide (and other greenhouse gases) in the atmosphere. That this might happen was first recognized in 1895 by Svante Arrhenius, a Swedish chemist and winner of the 1903 Nobel Prize in chemistry. Arrhenius realized that burning coal releases carbon dioxide into the atmosphere, and he estimated that a two- to threefold increase in carbon dioxide would cause a large temperature rise in northern regions, perhaps by as much as fifteen degrees Fahrenheit. Arrhenius thought this might make Stockholm suitably toasty, but he calculated that it would take several thousand years at the then-current rate of coal burning—far too long for him to enjoy the consequences. Arrhenius could not have anticipated how much coal—and also oil and gas—would be burned in the twentieth and twenty-first centuries, however. Over the past few hundred years, the carbon dioxide content of the atmosphere has risen by more than a third due to human activity, overwhelming the natural processes that had kept it roughly in balance over thousands of years before then. Finally, there is a third indisputable fact: through their decoding of the record preserved in rocks, ocean sediments, and ice cores, geoscientists have learned that in the past, high atmospheric carbon dioxide has characterized warm periods and low carbon dioxide has coincided with cold intervals. Even leaving aside the complexities of the climate system, or questions of positive and negative feedback, or the multiple uncertainties involved in modeling future climates, these three facts provide a compelling argument for taking very seriously the consensus view of climate scientists that greenhouse gases released by human activity are sharply raising global temperatures, and will continue to do so into the foreseeable future.

The role geoscientists play in this issue is to monitor, analyze, and model the climate system, past and present, and to forecast the probable consequences of different actions society may take in the face of rising greenhouse gas concentrations. Forecasting is a crucial but problematic part of this effort, because all forecasts, from estimating the global temperature one hundred years from now to predicting who will win next year's Superbowl, are uncertain. The key to accurate prediction is to reduce the uncertainty of the estimate to the lowest possible level. In the case of climate, this means that all known factors influencing climate, from surface albedo to cloudiness and ocean circulation, must be incorporated into the models; that both time and geographical resolution in the models must be high so that changes can be tracked over short time intervals and small patches of the Earth's surface; and that the causes and nature of past climate variation must be well understood. Fortunately, great progress has been made in all of these areas in recent decades, aided by the availability of large amounts of computing power. In spite of what you may sometimes hear in the media, the accuracy of climate models is rapidly improving, and confidence in their predictions is increasing. The difficulty lies in convincing policy makers and the public that the consequences of the predicted temperature increase demand present-day action, a problem compounded by the global scope of the phenomenon.

The course of ocean acidification is much easier to predict than is climate change, because the process is straightforward: as atmospheric carbon dioxide increases, more of the gas dissolves in seawater, making it more acidic. The solubility is well known, so the increase in acidity for a given increase in carbon dioxide can be calculated fairly readily. The difficulty—not for scientists, but for the ocean—is that the current rate of carbon dioxide increase is unprecedented. Natural feedback processes that act to decrease atmospheric carbon dioxide, such as increased weathering of continental rocks, or neutralization of ocean acidity by dissolution of sedimentary calcium carbonate, work on timescales that are long compared to the few centuries over which humans will add massive amounts of carbon dioxide to the atmosphere-ocean system (right now we add more than a million tons of carbon dioxide to the atmosphere *every hour*). The oceans themselves mix slowly compared to this timescale, which means that in the short term the acidification will mainly affect surface water in direct contact with the atmosphere (oceanographers have detected a measurable acidity increase in surface water over just the past two decades; since the Industrial Revolution, the increase has been about 30 percent). It will take thousands of years for the acidity to work its way through the entire ocean.

Perhaps because the focus of concern about rising greenhouse gases has been on global warming, ocean acidification has not received widespread attention until recently, and research into its effects is still in its early stages. However, national and international programs to study and monitor this phenomenon are now under way, and it is undoubtedly a topic that will increasingly occupy the spotlight. What has been discovered to date is not reassuring. Laboratory studies, for example, have already shown that reproduction in some marine species slows with increasing seawater acidity. Those affected include commercially important species such as oysters. Most marine organisms that build shells and skeletons from calcium carbonate will be negatively affected; they will be less able to make strong supporting and protective structures. Crucial among these are corals; among other things, coral reefs provide protection from waves and storm surges, and are important ecosystems for a wide range of marine life, including economically valuable fisheries.

These are not pleasant facts to learn, and I don't want to end this book on a pessimistic note. Problems like global warming and ocean acidification can sometimes seem so overwhelming and so unresponsive to individual actions that there is a temptation to simply throw up our hands and ignore them. But recognition of the problem is the first step toward solving it, and public opinion counts. That, I think, is shown by the actions of governments and organizations around the world, which—in spite of special interests and economic pressures—are now taking tentative steps aimed at maintaining a more stable surface environment on the Earth. Some of these steps would have been unimaginable just twenty-five or thirty years ago. The progress often seems agonizingly slow, but it is nevertheless progress. If the Anthropocene is the epoch when humans became a geological force, it also has the potential to be a time when human ingenuity is used to reverse at least some of the negative effects this development has brought. The only way that can be done is with a thorough understanding of how our planet works today, and how it has evolved through its long history. That message is loud and clear, and that is why the geosciences really do matter.

BIBLIOGRAPHY AND FURTHER READING

CHAPTER I

Bowring, S., D.H. Irwin, Y.G. Lin, M.W. Martin, K. Davidek, and W. Wang. 1998. U/Pb Zircon Chronology and Tempo of the End-Permian Mass Extinction. *Science* 280:1039–45.

Cutler, Alan. 2003. *The Seashell on the Mountaintop: A Story of Science, Sainthood, and the Humble Genius Who Discovered a New History of the Earth.* New York: Dutton.

Gradstein, Felix M., James G. Ogg, and Alan G. Smith, eds. 2005. *A Geologic Time Scale 2004.* Cambridge: Cambridge University Press. [Versions of the timescale from this book are available on the Internet, including on Wikipedia.]

Lyell, Charles. 1830–33. *Principles of Geology, Being an Attempt to Explain the Former Changes of the Earth's Surface, by Reference to Causes Now in Operation.* 3 vols. London: John Murray. [Available online at http://darwin-online.org.uk.]

Macdougall, Doug. 2008. *Nature's Clocks: How Scientists Measure the Age of Almost Everything.* Berkeley: University of California Press.

Repcheck, Jack. 2003. *The Man Who Found Time: James Hutton and the Discovery of the Earth's Antiquity.* Cambridge, MA: Perseus.

Urey, H.C., H.A. Lowenstam, S. Epstein, and C.R. McKinney. 1951. Measurement of Paleotemperatures and Temperatures of the Upper Cretaceous of England, Denmark, and the Southeastern United States. *Bulletin of the Geological Society of America* 62:399–416.

Winchester, Simon. 2001. *The Map that Changed the World: William Smith and the Birth of Modern Geology.* New York: HarperCollins.

CHAPTER 2

Cassidy, William A. 2003. *Meteorites, Ice and Antarctica: A Personal Account.* New York: Cambridge University Press.

Davis, Andrew M., ed. 2006. *Meteorites, Comets, and Planets: Treatise on Geochemistry, Vol. 1.* Oxford: Elsevier.

Hartmann, W.K., and D.R. Davis. 1975. Satellite-Sized Planetesimals and Lunar Origin. *Icarus* 24:504–15.

Huntington, Patricia A.M. 2002. Robert E. Peary and the Cape York Meteorites. *Polar Geography* 26:53–65.

Warren, P.H. 1985. The Magma Ocean Concept and Lunar Evolution. *Annual Review of Earth and Planetary Science* 13:201–40.

CHAPTER 3

Alvarez, L.W., W. Alvarez, F. Asaro, and H.V. Michel. 1980. Extraterrestrial Cause for the Cretaceous-Tertiary Extinction. *Science* 208:1095–1108.

Bottke, W.F., D. Vokrouhlický, and D. Nesvorný. 2007. An Asteroid Breakup 160 Myr Ago as the Probable Source of the K/T Impactor. *Nature* 449:48–53.

Buffetaut, Eric, and Christian Koeberl, eds. 2002. *Geological and Biological Effects of Impact Events.* Berlin: Springer.

Chapman, C.R. 2004. The Hazard of Near-Earth Asteroid Impacts on Earth. *Earth and Planetary Science Letters* 222:1–15.

Engledew, John. 2010. *The Tungus Event, or: The Great Siberian Meteorite.* New York: Algora.

Hayabusa at Asteroid Itokawa. 2006. Special issue, *Science* 312:1327–53.

Hildebrand, A.R., G.T. Penfield, D.A. Kring, M. Pilkington, A. Camargo Z., S.B. Jacobsen, and W.V. Boynton. 1991. Chicxulub Crater: A Possible Cretaceous/Tertiary Boundary Impact Crater on the Yucatán Peninsula, Mexico. *Geology* 19:867–71.

Hoyt, William Graves. 1987. *Coon Mountain Controversies: Meteor Crater and the Development of Impact Theory.* Tucson: University of Arizona Press.

Kring, D.A. 2007. The Chicxulub Impact Event and Its Environmental Consequences at the Cretaceous-Tertiary Boundary. *Palaeogeography, Palaeoclimatology, Palaeoecology* 255:4–21.

NASA. 2002. Hitchhiker's Guide to an Asteroid. *NASA Science News,* April 5, http://science.nasa.gov/science-news/science-at-nasa/2002/05apr_hitchhiker/.

NASA Near Earth Object Program Web site. http://neo.jpl.nasa.gov/neo/.

National Research Council. 2010. *Defending Planet Earth: Near-Earth Object Survey and Hazard Mitigation Strategies: Final Report.* Washington, DC: National Academies Press. [Available online at www.nap.edu.]

Nesvorný, D., D. Vokrouhlický, W. F. Bottke, B. Gladman, and T. Häggström. 2007. Express Delivery of Fossil Meteorites from the Inner Asteroid Belt to Sweden. *Icarus* 188:400–413.

Pilkington, M., A. R. Hildebrand, and C. Ortiz-Aleman. 1994. Gravity and Magnetic Field Modeling and Structure of the Chicxulub Crater, Mexico. *Journal of Geophysical Research* 99;13147–62.

Schmitz, B., M. Tassinari, and B. Peucker-Ehrenbrink. 2001. A Rain of Ordinary Chondritic Meteorites in the Early Ordovician. *Earth and Planetary Science Letters* 194:1–15.

Schulte, P., L. Alegret, I. Arenillas, J. A. Arz, P. J. Barton, P. R. Brown, T. J. Bralower, et al. 2010. The Chicxulub Asteroid Impact and Mass Extinction at the Cretaceous-Paleogene Boundary. *Science* 327:214–18.

University of New Brunswick Earth Impact Database Web site. www.unb.ca/passc/ImpactDatabase/.

CHAPTER 4

Allwood, A. C., M. R. Walter, B. S. Kamber, C. P. Marshall, and I. W. Burch. 2006. Stromatolite Reef from the Early Archaean Era of Australia. *Nature* 441:714–18.

Canfield, D. E. 2005. The Early History of Atmospheric Oxygen. *Annual Reviews of Earth and Planetary Science* 33:1–36.

Farquhar, J., and B. A. Wing. 2003. Multiple Sulfur Isotopes and the Evolution of the Atmosphere. *Earth and Planetary Science Letters* 213:1–13.

Haqq-Misra, J. D., S. D. Domagal-Goldman, P. J. Kasting, and J. F. Kasting. 2008. A Revised, Hazy Methane Greenhouse for the Archean Earth. *Astrobiology* 8:1127–37.

Nemchin, A., N. Timms, R. Pidgeon, T. Geisler, S. Reddy, and C. Meyer. 2009. Timing of Crystallization of the Lunar Magma Ocean Constrained by the Oldest Zircon. *Nature Geoscience* 2:133–36.

O'Neil, J., R. W. Carlson, D. Francis, and R. K. Stevenson. 2008. Neodymium-142 Evidence for Hadean Mafic Crust. *Science* 321:1828–31.

Sagan, C., and G. Mullen. 1972. Earth and Mars: Evolution of Atmospheres and Surface Temperatures. *Science* 177:52–56.

Wilde, S. A., J. W. Valley, W. H. Peck, and C. M. Graham. 2001. Evidence from Detrital Zircons for the Existence of Continental Crust and Oceans on the Earth 4.4 Gyr Ago. *Nature* 409:175–78.

CHAPTER 5

Burke, K. C., and J. T. Wilson. 1976. Hot Spots on the Earth's Surface. *Scientific American* 235:46–57.

Chen, C., S. Rondenay, R. L. Evans, and D. B. Snyder. 2009. Geophysical Detection of Relict Metasomatism from an Archean (~3.5 Ga) Subduction Zone. *Science* 326:1089–91.

Cox, Alan, and Robert Brian Hart. 1991. *Plate Tectonics: How It Works.* Hoboken, NJ: Wiley-Blackwell.

McPhee, John. 1998. *Annals of the Former World.* New York: Farrar, Straus and Giroux.

Oreskes, Naomi, ed. 2003. *Plate Tectonics: An Insider's History of the Modern Theory of the Earth.* Boulder, CO: Westview Press.

Vine, F. J., and D. H. Matthews. 1963. Magnetic Anomalies over Ocean Ridges. *Nature* 199:947–49.

CHAPTER 6

Atwater, B. F., M. Satoko, S. Kenji, T. Yoshinobu, U. Kazue, and D. K. Yamaguchi. 2005. *The Orphan Tsunami of 1700: Japanese Clues to a Parent Earthquake in North America.* U.S. Geological Survey Professional Paper 1707. Reston, VA: U.S. Geological Survey, in association with University of Washington Press. [Available online at http://pubs.usgs.gov/pp/pp1707/.]

Bell, J. W., J. N. Brune, L. Tanzhuo, M. Zreda, and J. C. Yount. 1998. Dating Precariously Balanced Rocks in Seismically Active Parts of California and Nevada. *Geology* 26:495–98.

Brune, J. N. 1996. Precariously Balanced Rocks and Ground-Motion Maps for Southern California. *Seismological Society of America Bulletin* 86:43–54.

Burchfiel, B. C., L. H. Royden, R. D. van der Hilst, B. H. Hager, Z. Chen, R. W. King, C. Li, J. Lü, H. Yao, and E. Kirby. 2008. A Geological and Geophysical Context for the Wenchuan Earthquake of 12 May 2008, Sichuan, People's Republic of China. *GSA Today* 18:4–11.

Johnston, A.C., and E.S. Schweig. 1996. The Enigma of the New Madrid Earthquakes of 1811–1812. *Annual Reviews of Earth and Planetary Science* 24:339–84.
U.S. Geological Survey Earthquakes Web site. http://earthquake.usgs.gov/earthquakes/.
Van Arsdale, R.B., and R.T. Cox. 2007. The Mississippi's Curious Origins. *Scientific American* 296 (1):76–82B.
Witze, A. 2009. The Sleeping Dragon. *Nature* 459:153–57.

CHAPTER 7

Campbell, I.H., and C.M. Allen. 2008. Formation of Supercontinents Linked to Increases in Atmospheric Oxygen. *Nature Geoscience* 1:554–58.
Canfield, D.E., S.W. Poulton, and G.M. Narbonne. 2007. Late-Neoproterozoic Deep-Ocean Oxygenation and the Rise of Animal Life. *Science* 315:92–95.
Hoffman, P.F., A.J. Kaufman, G.P. Halverson, and D.P. Schrag. 1998. A Neoproterozoic Snowball Earth. *Science* 281:1342–46.
Kirschvink, J.L. 1992. Proterozoic Low-Latitude Global Glaciation: The Snowball Earth. In *The Proterozoic Biosphere*, ed. J. William Schopf and Cornelis Klein, 51–52. Cambridge: Cambridge University Press.
Kump, L.R. 2008. The Rise of Atmospheric Oxygen. *Nature* 451:277–78.
Narbonne, G.M., and J.G. Gehling. 2003. Life after Snowball: The Oldest Complex Ediacaran Fossils. *Geology* 31:27–30.
Reddy, S.M., R. Mazumder, D.A.D. Evans, and A.S. Collins, eds. 2009. Palaeoproterozoic Supercontinents and Global Evolution. *Geological Society Special Publication 322*. London: Geological Society.
Rogers, J.J.W., and M. Santosh. 2003. Supercontinents in Earth History. *Gondwana Research* 6:357–68.
———. 2009. Tectonics and Surface Effects of the Supercontinent Columbia. *Gondwana Research* 15:373–80.
Torsvik, T.H. 2003. The Rodinia Jigsaw Puzzle. *Science* 300:1379–81.
Walker, Gabrielle. 2003. *Snowball Earth: The Story of the Great Global Catastrophe that Spawned Life as We Know It*. New York: Crown Publishers.
Zimmer, C. 2009. On the Origin of Eukaryotes. *Science* 325:666–68.

CHAPTER 8

Alley, Richard B. 2000. *The Two-Mile Time Machine: Ice Cores, Abrupt Climate Change, and Our Future*. Princeton: Princeton University Press.

Andersen, K.K., N. Azuma, J.-M. Barnola, M. Bigler, P. Biscaye, N. Caillon, J. Chappellaz, et al. 2004. High-Resolution Record of Northern Hemisphere Climate Extending into the Last Interglacial Period. *Nature* 431:147–51.

Cox, P., and C. Jones. 2008. Illuminating the Modern Dance of Climate and CO_2. *Science* 321:1642–43.

Dansgaard, W. 2000. *Frozen Annals: Greenland Ice Sheet Research*. Copenhagen: Niels Bohr Institute, University of Copenhagen. [Available online at www.iceandclimate.nbi.ku.dk/publications/frozen_annals/.]

Huybers, P. 2009. Antarctica's Orbital Beat. *Science* 325:1085–86.

Jansen, E., J. Overpeck, K.R. Briffa, J.-C. Duplessy, F. Joos, V. Masson-Delmotte, D. Olgao, et al. 2007. Palaeoclimate. In *Climate Change 2007: The Physical Science Basis. Contribution of Working Group I to the Fourth Assessment Report of the Intergovernmental Panel on Climate Change,* ed. S. Solomon, D. Qin, M. Manning, Z. Chen, M. Marquis, K.B. Averyt, M. Tignor, and H.L. Miller. Cambridge: Cambridge University Press. [Available online at www.ipcc.ch/publications_and_data/publications_and_data_reports.htm#2.]

Jouzel, J., et al. 2007. Orbital and Millennial Antarctic Climate Variability over the Past 800,000 Years. *Science* 317:793–96.

Langway, C.C., Jr. 2008. *The History of Early Polar Ice Cores*. ERDC/CRREL TR-08–1. Hanover, NH: U.S. Army Corps of Engineers, Engineer Research and Development Center. [Available online at www.crrel.usace.army.mil/library/technicalpublications.html.]

Lurie, Edward. 1960. *Louis Agassiz: A Life in Science*. Chicago: University of Chicago Press.

Macdougall, Doug. 2004. *Frozen Earth: The Once and Future Story of Ice Ages*. Berkeley: University of California Press.

Muller, Richard A., and MacDonald, Gordon J. 2000. *Ice Ages and Astronomical Causes*. Chichester: Springer-Praxis.

Murton, J.B., M.D. Bateman, S.R. Dallimore, J.T. Teller, and Z. Yang. 2010. Identification of Younger Dryas Outburst Flood Path from Lake Agassiz to the Arctic Ocean. *Nature* 464:740–43.

Petit, J.R., J. Jouzel, D. Raynaud, N.I. Barkov, J.M. Barnola, I. Basile, M. Bender, et al. 1999. Climate and Atmospheric History of the Past 420,000 Years from the Vostok Ice Core, Antarctica. *Nature* 339:429–36.

Raymo, M.E., and P. Huybers. 2008. Unlocking the Mysteries of the Ice Ages. *Nature* 451:284–85.

CHAPTER 9

Beerling, D.J. 2009. Enigmatic Earth. *Nature Geoscience* 2:537–38.

Kennett, J.P., and L.D. Stott. 1991. Abrupt Deep-Sea Warming, Palaeoceanographic Changes and Benthic Extinctions at the End of the Palaeocene. *Nature* 353:225–29.

Nisbet, E.G., S.M. Jones, J. Maclennan, G. Eagles, J. Moed, N. Warwick, S. Bekki, P. Braesicke, J.A. Pyle, and C.M.R. Fowler. 2009. Kick-starting Ancient Warming. *Nature Geoscience* 2:156–58.

Nunes, F., and R.D. Norris. 2006. Abrupt Reversal in Ocean Overturning during the Palaeocene/Eocene Warm Period. *Nature* 439:60–63.

Wing, S.L., G.J. Harrington, F.A. Smith, J.I. Bloch, D.M. Boyer, and K.H. Freeman. 2005. Transient Floral Change and Rapid Global Warming at the Paleocene-Eocene Boundary. *Science* 310:993–96.

Zachos, J.C., U. Röhl, S.A. Schellenberg, A. Sluijs, D.A. Hodell, D.C. Kelly, E. Thomas, et al. 2005. Rapid Acidification of the Ocean during the Paleocene-Eocene Thermal Maximum. *Science* 308:1611–15.

Zeebe, R.E., J.C. Zachos, and G.R. Dickens. 2009. Carbon Dioxide Forcing Alone Insufficient to Explain Palaeocene-Eocene Thermal Maximum Warming. *Nature Geoscience* 2:576–80.

CHAPTER 10

Bralower, T.J. 2008. Volcanic Cause of Catastrophe. *Nature* 454:285–87.

Hallam, Tony. 2005. *Catastrophes and Lesser Calamities: The Causes of Mass Extinctions*. Oxford: Oxford University Press.

Kump, L.R., A. Pavlov, and M.A. Arthur. 2005. Massive Release of Hydrogen Sulfide to the Surface Ocean and Atmosphere during Intervals of Oceanic Anoxia. *Geology* 33:397–400.

Mahoney, John J., and Millard F. Coffin, eds. 1997. *Large Igneous Provinces: Continental, Oceanic, and Planetary Flood Volcanism*. Washington, DC: American Geophysical Union.

Turgeon, S.C., and R.A. Creaser. 2008. Cretaceous Oceanic Anoxic Event 2 Triggered by a Massive Magmatic Episode. *Nature* 454:323–25.

Visscher, H., C.V. Looy, M.E. Collinson, H. Brinkhuis, J.H.A. van Konijnenburg-van Cittert, W.M. Kürschner, and M.A. Sephton. 2004. Environmental Mutagenesis during the End-Permian Ecological Crisis. *Proceedings of the National*

Academy of Sciences of the United States of America 101:12952–56. [Available online at www.pnas.org.]

Ward, P. D. 2006. Impact from the Deep. *Scientific American* 295 (4):64–71.

CHAPTER 11

Ambrose, S. H. 1998. Late Pleistocene Human Population Bottlenecks, Volcanic Winter, and Differentiation of Modern Humans. *Journal of Human Evolution* 34:623–51.

Ashfall Fossil Beds State Historical Park Web site. www.ashfall.unl.edu/.

Bindeman, I. N. 2006. The Secrets of Supervolcanoes. *Scientific American* 294 (6):36–43.

Mason, B. G., D. M. Pyle, and C. Oppenheimer. 2004. The Size and Frequency of the Largest Explosive Eruptions on Earth. *Bulletin of Volcanology* 66:735–48.

Sparks, S., S. Self, et al. 2005. *Super-eruptions: Global Effects and Future Threats.* Report of a Geological Society of London Working Group. [Available online at www.geolsoc.org.uk/supereruptions.]

Stommel, Henry, and Elizabeth Stommel. 1983. *Volcano Weather: The Story of 1816, the Year without a Summer.* Newport, RI: Seven Seas Press.

U.S. Geological Survey Mount Pinatubo resources page. http://vulcan.wr.usgs .gov/Volcanoes/Philippines/Pinatubo/framework.html.

U.S. Geological Survey Volcanic Hazards Program Web site. http://volcanoes .usgs.gov/.

U.S. Geological Survey Yellowstone Volcano Observatory Web site. http:// volcanoes.usgs.gov/yvo/.

Wark, D. A., and C. F. Miller, eds. 2008. Supervolcanoes. Special issue, *Elements* 4:11–48.

CHAPTER 12

Ahlberg, P. E., and J. A. Clack. 2006. A Firm Step from Water to Land. *Nature* 440:747–48.

Bambach, R. K. 2006. Phanerozoic Biodiversity Mass Extinctions. *Annual Review of Earth and Planetary Science* 34:127–35.

Clack, Jennifer A. 2002. *Gaining Ground: The Origin and Early Evolution of Tetrapods.* Bloomington: Indiana University Press.

———. 2005. Getting a Leg Up on Land. *Scientific American* 293 (6):100–107.

Engel, M. S., and D. A. Grimaldi. 2004. New Light Shed on the Oldest Insect. *Nature* 427:627–30.

Gould, Stephen Jay. 1989. *Wonderful Life: The Burgess Shale and the Nature of History.* New York: W. W. Norton.

Jackson, J. B. C. 2008. Ecological Extinction and Evolution in the Brave New Ocean. *Proceedings of the National Academy of Sciences of the United States of America* 105:11458–65. [Available online at www.pnas.org.]

Kenrick, P. 2003. Fishing for the First Plants. *Nature* 425:248–49.

Ward, Peter Douglas. 2007. *Under a Green Sky: Global Warming, the Mass Extinctions of the Past, and What They Can Tell Us about Our Future.* New York: Smithsonian Books/Collins.

Wyss, A. 2001. Digging Up Fresh Clues about the Origin of Mammals. *Science* 292:1496–97.

CHAPTER 13

Hazen, R. M., D. Papineau, W. Bleeker, R. T. Downs, J. M. Ferry, T. J. McCoy, D. A. Sverjensky, and H. Yang. 2008. Mineral Evolution. *American Mineralogist* 93:1693–1720.

Inman, M. 2009. A Sensitive Subject. *Nature Reports Climate Change,* April 30. www.nature.com/climate/2009/0905/full/climate.2009.41.html.

NSF Advisory Committee for Geosciences. 2009. *GEO Vision: Unraveling Earth's Complexities through the Geosciences.* [Available online at www.nsf.gov/geo/acgeo/geovision/start.jsp.]

Rockström, J., W. Steffen, K. Noone, Å. Persson, F. S. Chapin III, E. F. Lambin, T. M. Lentin, et al. 2009. A Safe Operating Space for Humanity. *Nature* 461:472–75.

U.S. Geological Survey Mineral Resources Program Web site. http://minerals .usgs.gov/.

U.S. Geological Survey Water Resources of the United Web site. http://water .usgs.gov/.

INDEX

Page numbers in italic refer to illustrations.

Text:	10.75/15 Janson
Display:	Janson MT Pro
Compositor:	BookMatters, Berkeley
Printer and binder:	Thomson-Shore, Inc.

Frozen Earth
The Once and Future Story of Ice Ages

"Should be required reading for anyone interested in the future of the planet."
—*Times Higher Education Supplement*

"A comfortable-to-read ecotale. ... Macdougall's writing is easy, pegging narrative to history of discovery."—*Nature*

"Packed with detailed information. ... Macdougall's engaging style makes it a pleasurable and thought-provoking read."
—*Library Journal*

$19.95 paper 978-0-520-24824-3

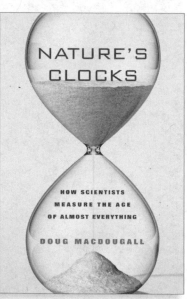

Nature's Clocks
How Scientists Measure the Age of Almost Everything

"For time-conscious readers, *Nature's Clocks* provides satisfaction beyond measure."—*Washington Post Book World*

"Rich in historical tidbits, this book is a delightful study of how scientists figured out analytical techniques that revealed the history of the earth."—*New Scientist*

"Science buffs from all fields along with general readers will find this a helpful handbook on how we are now able to travel to the distant past."—*Publishers Weekly*

$17.95 paper 978-0-520-26161-7

www.ucpress.edu